On Beyond Euclid

In Which We Take a Guided Tour Through Book 1 of Euclid's <u>Elements</u>, Whilst Being Continually Distracted by Einstein's Special Theory of Relativity and Other Pesky Non-Euclidean Diversions

On Beyond Euclid

In Which We Take a Guided Tour Through Book I of Euclid's <u>Elements</u>, Whilst Being Continually Distracted by Einstein's Special Theory of Relativity and Other Pesky Non-Euclidean Diversions

by Ben Jacobs

"From there to here,
from here to there,
funny things
are everywhere."
--Dr. Seuss

To my family,
for keeping me firmly planted
in this universe.

Table of Contents

Introduction

This curious little book is basically a hitchhiker's guide to Book I of Euclid's *Elements*. We travel through each of the forty-eight Propositions—more or less in order—and see how each one generalizes (or does not generalize) to hyperbolic and other non-Euclidean spaces. The prerequisite is high-school (circular) trigonometry.

Few people seem to realize that Einstein's special theory of relativity is a model of hyperbolic geometry. See the table below.

Glossary	Minkowski Geometry	Hyperbolic Geometry
Point	Event in space-time	Particle in uniform motion
Distance between two points	Space-time interval between two events	Relative speed between two such particles

While the connection between Minkowski geometry and special relativity is *well-known*, the connection between hyperbolic geometry and special relativity is, rather, *known of.* But this book makes the hyperbolic connection explicit; I use Poincaré disks rather than Minkowski diagrams to illustrate concepts of special relativity. I think this is perhaps the only accessible book in the English language to do so. Moreover, relativity is introduced fairly early (Task 10 of Universe 1) and is continually referred to. (There are some *advanced* books in differential geometry that use relativity as a motif, notably the wonderful *Semi-Riemannian Geometry* by Barrett O'Neill. However, the prerequisite for *his* book is basically an undergraduate degree in mathematics.)

If you happen to be a physicist reading this Introduction, you may well find yourself, at this point, somewhat bewildered, as though you were walking down a familiar street, but all of the houses were painted purple. (If you're not a physicist, you may safely skip to the next paragraph.) As you well know, *Minkowski geometry,* not hyperbolic geometry, is how (most) physicists visualize the concepts of special relativity, and with good reason. Some of the most counterintuitive ideas (such as the relativity of simultaneity) are particularly well suited to Minkowski diagrams, and you can be forgiven for not seeing the need to try anything else. However, trust me; there *are* good reasons. As a simple example, consider the following exercise: "*A* thinks *B* is going 40% of the speed of light, and *B* thinks *C* is going 50% of the speed of light. So *C* must think that *A* is going between --?--% of the speed of light and --?--% of the speed of light." In hyperbolic geometry this is a straightforward triangle-inequality problem on the Poincaré disk, and even weaker students find it easily doable: no fancy formulas needed. And that's just the beginning; there's more inside. Indeed, *every theorem in hyperbolic geometry is a theorem in relativistic kinematics, and conversely.* Now I'm not at all arguing that the Poincaré disks are *better* than Minkowski diagrams. I'm just saying that they *do* have their place. It's another arrow in your quiver, so why not use it? Also, from a purely *mathematical* point of view, hyperbolic geometry is far richer and more interesting than Minkowski geometry, which isn't even a metric space. (Actually, since you're a physicist, that last sentence probably isn't compelling. Oh well...) Now, back to the Introduction...

A major inspiration for me in writing this was *The Linear Algebra Problem Book* by Paul Halmos. This book, like his, is problem-driven.

This book has a lot of proofs in it, but a unique feature is that virtually all of the proofs are in the form of fill-in-the-blank exercises. It seemed to me that making the students do these proofs from scratch would be unreasonable, but simply including the proofs as part of the text would make them mostly unread. But I wanted the proofs to *be* read, for they are an essential part of the course. The fill-in-the-blank approach seemed to me to be Buddha's middle way.

When I teach non-Euclidean geometry, I usually use *Geometry, Relativity, and the Fourth Dimension* by Rudy Rucker as a supplemental text. Our books seem to complement each other; they overlap just enough so that students can see the connection, but not so much that they get overwhelmed with repetition. (Also, his book provides the necessary Minkowski diagrams that my book lacks.) As a reference, I always use Volume I of the Sir Thomas Heath translation of *The Thirteen Books of The Elements* by Euclid, available from Dover Press: <http://www.doverpublications.com/>.

I fell in love with geometry after reading Marvin Jay Greenberg's *Euclidean and Non-Euclidean Geometries.* Indeed, the first chance I got, I used Greenberg's book as the text for my course, but it proved to be too advanced for my students. I searched around for

Introduction

a more appropriate book, and since I could not find the one I was looking for, I had to write it myself.

A solutions manual is available, for teachers and autodidacts. Contact me at ben.jacobs@sfuhs.org for more information.

Acknowledgements

I must thank my students at San Francisco University High School who took my course *Non-Euclidean Geometry, Relativity, and the Fourth Dimension*; they suffered through preliminary drafts of this text, and their complaints (and occasional praise) were invaluable.

I want to thank my math colleagues at University High for allowing me the freedom to teach such a nonstandard course in the first place. I also want to thank Michael Diamonti for first proposing the idea of a sabbatical program at our school, and the faculty at large for granting me one of the first semester sabbaticals, which gave me the opportunity to shepherd my scattered notes into some semblance of order.

Most of the illustrations in this book were made using the freeware NonEuclid written by Joel Castellanos, available at http://cs.unm.edu/~joel/NonEuclid/. A few of them were made using Geometers Sketchpad, available from Key Curriculum Press.

Lastly, but mostly, I thank my family, for their unwavering love and support.

Introduction

Prologue

*In Which We Briefly Review Some Undefinitions and Some Common Notions,
Before We Commence Our Grand Tour of Euclid's Elements*

A point is a point, a line is a line,
  a rose is a rose is a rose.
We thus undefine in the manner of Stein
  some terms in unrhyme and unprose.
On these as foundation we lay definitions,
  the girders for walls and a roof.
We assume some conditions to fit requisitions
  and build us a logical proof.
When Hilbert found flaws in the structural scene,
  he repaired the unrigor discreetly.
We know what we mean when we speak of "between,"
  which Euclid unnoticed completely.
As we trace mathematics up from its roots
  to the reach of its vast propositions,
the beautiful fruits of our lively pursuits
  all stem from our undefinitions.
--Katherine E. O'Brien

The Five Undefined Geometric Terms of Plane Geometry

point
line
lie on, lies on
congruent
between

Undefined Term	Context
point	A noun. "I have five points."
line	A noun. "I have seventeen lines."
lie on, lies on	Only undefined when referring to points lying on lines, e.g., "Point P lies on line l." In other contexts, such as "Point P lies on circle α," we need to define what we mean.
congruent	Only undefined when referring to segments or angles. In other contexts, such as "these two triangles are congruent," we need to define what we mean.
between	Refers to three collinear points, or three coterminal rays.

Other Math Words You Can Use in Your Definitions
set
element
union
intersection (two lines can *intersect* at a point P, which means (of course) that P *lies on* both lines)

Properties of Algebra

Reflexive Property   If $a = b$, then $b = a$. Also, $a = a$.

Addition Property   If $a = b$ and $c = d$, then $a + c = b + d$.

Multiplication Property   If $a = b$, then $ac = bc$

Substitution Property   If $a = b$, then a or b can be substituted
for   the other in any mathematical
expression.

Distributive Property $a(b + c) = ab + ac.$

                                          Prologue

Properties of Analysis

Addition Property — If $a > b$ and $c \geq d$, then $a + c > b + d$.

Multiplication Properties — If $c > 0$, and $a > b$, then $ac > bc$.

If $c < 0$, and $a > b$, then $ac < bc$.

Transitive Property — If $a > b$ and $b > c$, then $a > c$.

The Whole Equals the Sum of Its Parts (TWESP) — If $a = b + c$ and $c > 0$, then $a > b$.

Task 1: In this course we are going to precisely define lots of mathematical terms. In our definitions, we are going to use common and ubiquitous English words like *the*, *a*, and *and*, which we will never bother defining; we're not going to be fanatical here. We will also use previously defined words in our definitions, which is fine and makes sense. But here's the bizarre thing: we're *also* going to use words that we are intentionally *never* going to define. These terms are: *point*, *line*, *lie on*, *congruent*, and *between*. Now these terms will only be undefined in certain contexts. For example *congruent* is only an undefined word when referring to segments or angles; if we talk about *congruent circles*, we need to define what we mean. Moreover, to speed things up a bit, we will also allow the use of these common mathematical terms, that someday you may see rigorously defined in a more advanced course in Set Theory: *set*, *element*, *union*, and *intersection*. Even though we are never going to define these words, we will not hesitate to use them in our definitions of other words.

Which begs the question: why do we have undefined terms at all? Why don't we just define every word?

Task 2: See if you can fill in the blanks of the following "definitions," due to Euclid. (Hint: Euclid didn't know about undefined terms.)

(a) A ________ is that which has no part.

(b)   A _________ is a breadthless length.

(c)   A _________ is a curve which lies evenly with the
        points on itself.

(d)   A _________ is a plane figure contained by one curve
        such that all the line segments falling upon it from
        one point among those lying within the figure are
        congruent to one another.

Task 3: Referring to the previous Task, if you didn't have any idea
about what these words meant, would Euclid's definitions help?

Task 4: Besides definitions, Euclid's geometry book, *The Elements*,
also contained what he called "Common Notions," which today we
would call "Algebraic Properties" or "Axioms of Algebra." Euclid's
three Common Notions are:

(1)   Things that are equal to the same thing are also equal
        to each other.

(2)   If equals be added to equals, the wholes are equal.

(3)   If equals be subtracted from equals, the remainders
        are equal.

Which Properties of Algebra do each of these Common Notions
correspond to?

Task 5:   During the first quarter of the twentieth century,
probably the leading mathematician in the world was David Hilbert
(1862-1943). You may not have heard of him—which, if you think
about it, is not all that strange, for how many mathematicians *have*
you heard of?—but he discovered and created many wonderful
things mathematical. (No need to take *my* word for it; read the
fascinating biography of him by Constance Reid.) In 1900, at an
international gathering of mathematicians in Paris, he submitted 23
important problems to be targeted during the 20$^\text{th}$ century. They
were instantly a smash hit (among the mathematical crowd), and as of
this writing, many—but not all—of *Hilbert's problems* have been

solved. But solved or unsolved, they were and remain incredibly influential in the development of mathematics.

Of his many accomplishments, the one that concerns us here is his work in geometry. Hilbert went through all of the postulates and proofs in Euclid's *Elements* with a fine-toothed comb, and put them on a rigorous footing. Indeed, the list of axioms that we will develop in this book will loosely follow his system.

He was fond of saying: "One must be able to say at all times—instead of *points, lines,* and *planes—tables, chairs,* and *beer mugs.*" What do you think that he meant by this?

Universe 1: Neutral Euclid
In Which We Almost Prove Euclid's Proposition 1

> *For out of the old feldes, as men seith,*
> *Cometh al this new corne fro yeere to yere;*
> *And out of olde bokes, in good feith,*
> *Cometh al this new science that men lere.*
> *--Chaucer*

A Modern Rewording of Euclid's First Four Postulates

1. For every point P and for every point Q not equal to P there exists a unique line l that lies on P and Q.

2. Given segments $\overline{AB}$ and $\overline{CD}$, there exists a unique point E such that B is between A and E (or $A*B*E$), and $\overline{CD} \cong \overline{BE}$.

3. For every point O and every point A not equal to O there exists a circle with center O and radius $\overline{OA}$.

4. All right angles are congruent to each other.

Task 1: Our first universe consists of Euclid's first four postulates from Book I of *The Elements*. Euclid actually used *five* postulates in all, but we're omitting the last postulate for now. Here is the postulate that we're omitting:

5. For every line l and for every point P that does not lie on l, there exists a unique line m through P that is parallel to l.

In fact, that's why this universe is called "neutral." For we are not yet taking a stand about whether his last postulate is true or not. Indeed, Euclid's fifth postulate—the Parallel Postulate—was controversial from the beginning. Even the Very Wise thought that instead of being a postulate, it should be a *theorem*, that is, they thought that it could be proved from the other four postulates. So for more than 2000 years, the best mathematical minds strove in vain to find a *proof* for the Parallel Postulate. Their ultimate failure in the late nineteenth century shattered the conception of Euclidean geometry as the true description of physical space, and paved the way for Einstein's Theory of Relativity. Indeed, it turns out that all of the theorems in Neutral Geometry are true not only in Euclidean Geometry, but in Einstein's Special Theory of Relativity as well. We'll see the connection soon. But we digress...

Euclid wrote his book *The Elements* in ancient Greek. Here is a modern English translation (by Sir Thomas Heath) of the original wording of Euclid's first four postulates:

> *Let the following be postulated:*
> 1.   *To draw a straight line from any point to any point.*
> 2.   *To produce a finite straight line continuously in a straight line.*
> 3.   *To describe a circle with any centre and distance.*
> 4.   *That all right angles are equal to one another.*

Why do you suppose that this book (along with all other modern math books) alters the wording of these postulates, instead of just sticking to Euclid's original wording?

Task 2: Guess what? This book does not contain a glossary! That means that you are going to have to construct your own glossary, because you will need to invoke definitions when we do proofs. So let's start now! (Helen Keller once said, "Life is a daring adventure, or nothing!") Postulate 2 speaks of the "segments $\overline{AB}$ and $\overline{CD}$." But we have not yet defined what a segment is. Define *segment*. You may use any of the undefined terms in your definition. You may also use the words *set* and *element* in your definition. [Hint: I'll start you off. "Given two points A and B, the *segment* $\overline{AB}$ is..."] While you're at it, also define *endpoints*.

Task 3: There are creatures from another universe who are also reading this book. (Their universe is called *Poincaré Disk Space*, by the way.) They also just did Task 2, and they also got the same answer that you did. However, they visualize *lines* and *segments* differently than you do. (See Figure 1-1.) They think of the entire infinite

20

"plane" as a finite circular region. They think of "lines" as arcs of circles whose endpoints are perpendicular to the circle. Diameters that go through the center of the circle are also "lines." So the curves CD and EF in the picture are "lines," and AB is a "segment." They measure distance in a convoluted way, too (from our point of view). They think that the shortest "distance" between the two points A and B is along the curved "segment." Similarly, the shortest "distance" between points C and D, and between points E and F are along the "lines." All of this seems perfectly normal and obvious to them. The creatures are also reading this problem right now, but they see a different picture than you do. They see *your* version of a "plane," "lines," and "segment." They think that *your* lines are not straight. They think *you're* weird. Question: do you think they're weird?

Figure 1-1: *"Lines"* $\overleftrightarrow{CD}$ *and* $\overleftrightarrow{EF}$ *, and "segment"* $\overline{AB}$ *in Poincaré Disk Space.*

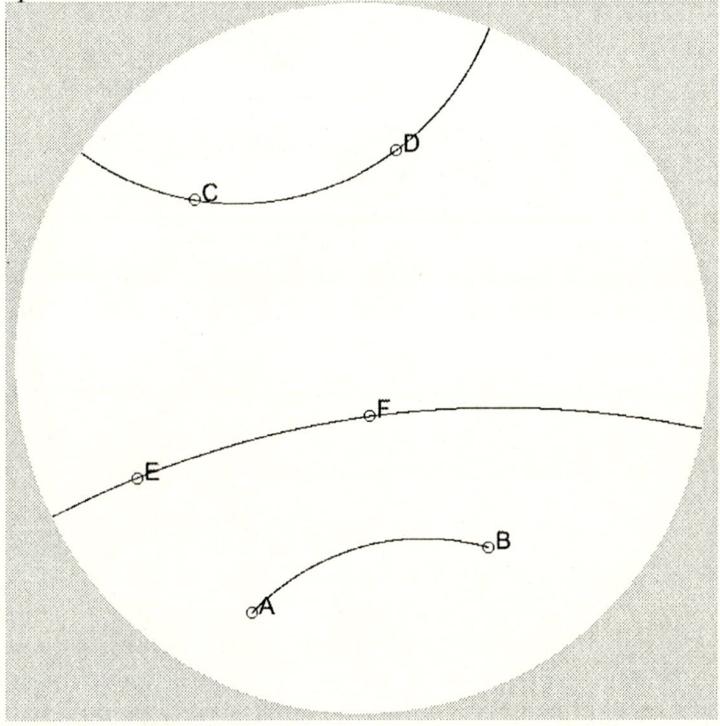

Task 4: Here's another word for your glossary! Define *collinear*. [Hint: "Points *A, B,* and *C* are *collinear* if..."] If it will help, the points

   Universe 1: Neutral Euclid

A, *B*, and *C* in Figure 1-2 are "collinear," from the creatures-in-the-other-universe's point of view.

Task 5: Yet another word for your glossary. Define *triangle*. [Hint: "Let *A, B,* and *C* be three non-collinear points. The triangle Δ*ABC* is..."] Also define *equilateral triangle*. While you're at it, also define *isosceles triangle*.

     If it will help, in Figure 1-3, Δ*ABD* is an equilateral "triangle." It doesn't look like an equilateral triangle—or even like an equilateral "triangle"—to us, because side *AD* and *BD* look longer than side *AB*. Perhaps it could be an isosceles "triangle," but it's definitely not equilateral.

Figure 1-2: Collinear points in Poincaré Disk Space

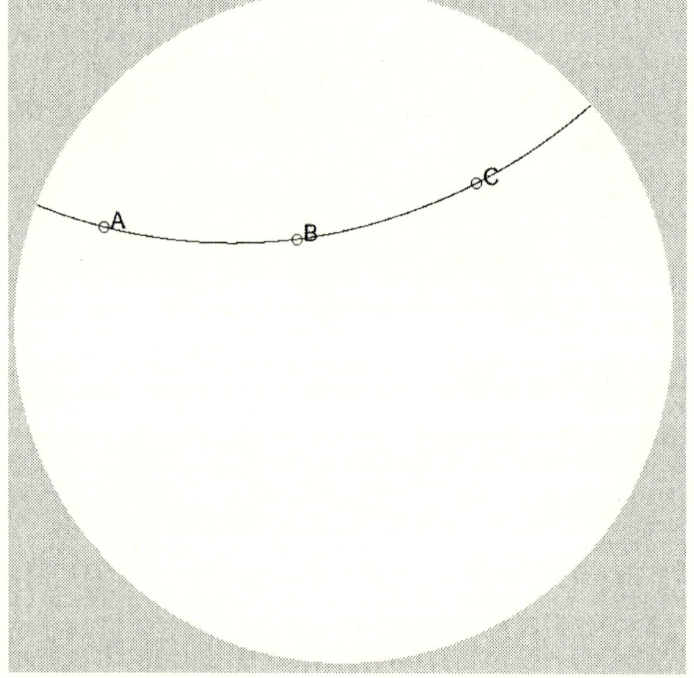

However, to the creatures-in-the-other-universe's eyes, all of the sides are the same length. (All the sides are also "straight," of course.) The creatures measure the sides of their triangles with rulers marked off in inches, just like we do, but from *our* point of view their rulers shrink as they approach the boundary of the circle. In fact, *everything*

Universe 1: Neutral Euclid

shrinks as it approaches the boundary of the circle. The creatures, of course, think no such thing. They think that their rulers (and they, themselves) are always the same size, no matter where they are in the circle. (For example, they think that all of the triangles in the "plane" in Figure 1-4 are congruent.) They also think that their "lines" are infinitely long, for if they tried to measure the length of a line with their "rulers," it would take forever (since the size of their rulers shrink to nothing as they get closer and closer to the boundary of the circle). In fact, they don't think that their plane has a boundary at all.

They're reading this right now, of course, but they see a different picture than you do. They see a picture of *your* plane: *the whole thing!* You may not quite understand how they can have a picture of the *whole* Euclidean plane in their books. After all, from your point of view, a plane goes on forever in two dimensions, so how could the picture of a whole plane fit on a page? It's a conundrum.

Figure 1-3: An equilateral triangle in Poincaré Disk Space

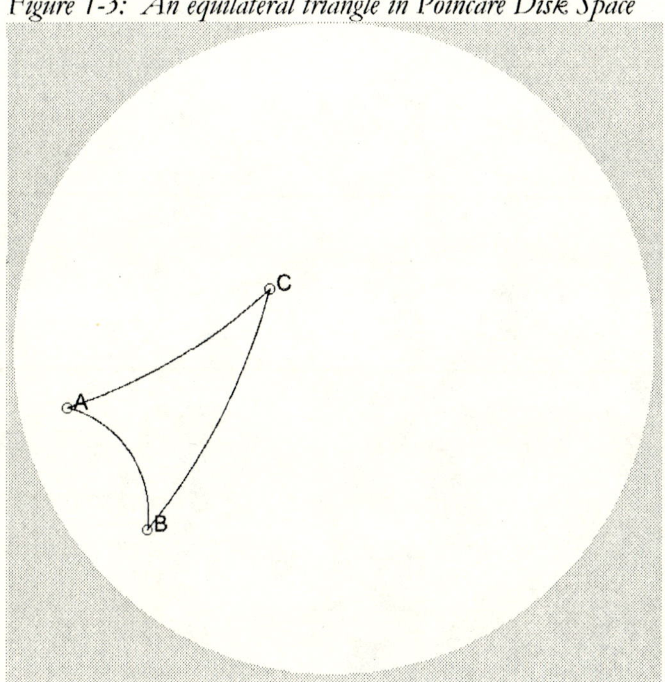

Task 6: More words for your glossary. Define *circle* with *center A* and *radius $\overline{AB}$*. [Note: If you define a circle as a kind of a set, then this renders Postulate 3 unnecessary, since by set theory such a circle would have to exist. (It may be the null set, but it still exists.) So ironically, for two thousand years, while mathematicians strove to prove that Eulcid's Postulate 5 was superfluous, they completely overlooked Postulate 3!]

Figure 1-5 shows a "circle" with center A and radius $\overline{AB}$. Notice that "circles" in the other universe look just like circles in this universe, except that the "center" A looks off-center. They, of course, have similar opinions about *our* circles. If you mark a point C on their "circle" at random, from their point of view the distance between A and C equals the distance between A and B. In Figure 1-6, the segments $\overline{AC}$ and $\overline{AB}$ both have the same "length." Indeed, they are both "radii."

Figure 1-4: Congruent triangles in Poincaré Disk Space

Universe 1: Neutral Euclid

Task 7: *Poincaré Disk Space* is named after the famous mathematician who created/discovered it, Henri Poincaré (1854-1912). We've already discussed this space, but let's take a closer look at it.

Figure 1-5: "Circle" with center A and radius AB.

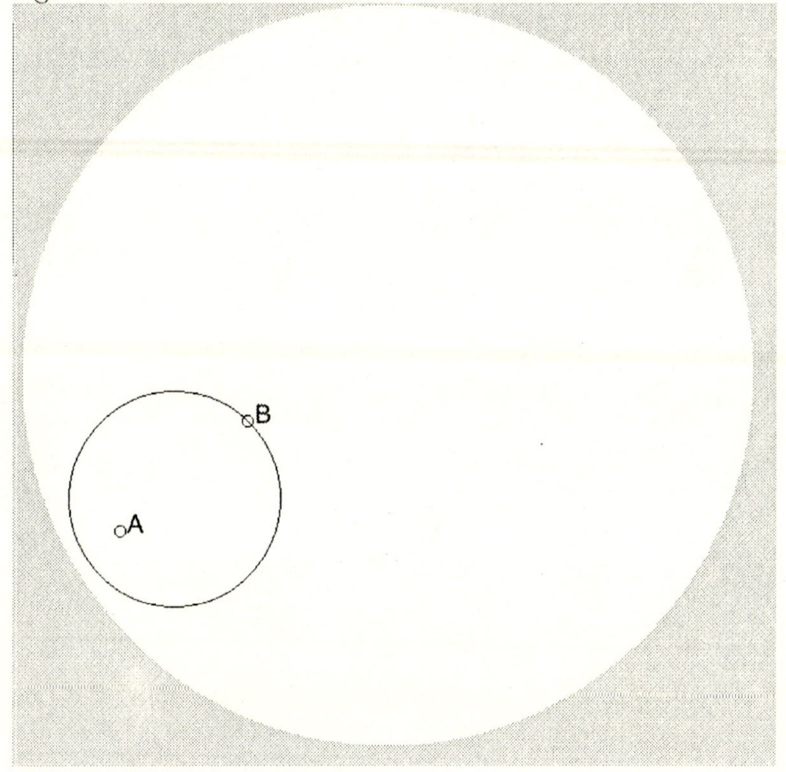

We can think of this universe as being the inside of the unit circle $x^2 + y^2 = 1$. Interpret "point" to mean an ordinary point inside the unit circle. Interpret "line" to mean either a diameter of the unit circle, or else a circular arc in the interior of the unit circle and *orthogonal to it.* (This last bit is important.) Let "lie on" mean that a "point" lies on a "line" in the usual way. In the math biz, this model is called *conformal,* which means that the "angle" between two Poincaré "lines" is the regular Euclidean angle between their tangents. This is good; a 30° angle *looks* like a 30° angle, even to our jaded Euclidean eyes. (No need to don our special glasses.) Again, the measurements of lengths needs a special hyperbolic "ruler" that shrinks to nothing as it approaches the boundary of the unit circle.

25

Flatlander who lives in Poincaré Disk Space would—from our perspective—appear to shrink as she approached the boundary of her universe: the unit circle. Again, from the Flatlander's perspective, she would *not* think that she was shrinking, but she *would* think that the boundary of her universe was infinitely far away, no matter how "close" she got to it! In the Escher drawing in Figure 1-7, all of the bats are congruent; the bats just appear to get smaller as they approach the boundary of the circle because we're looking at them through Euclidean eyes.

Figure 1-6: AC = AB in Poincaré Disk Space

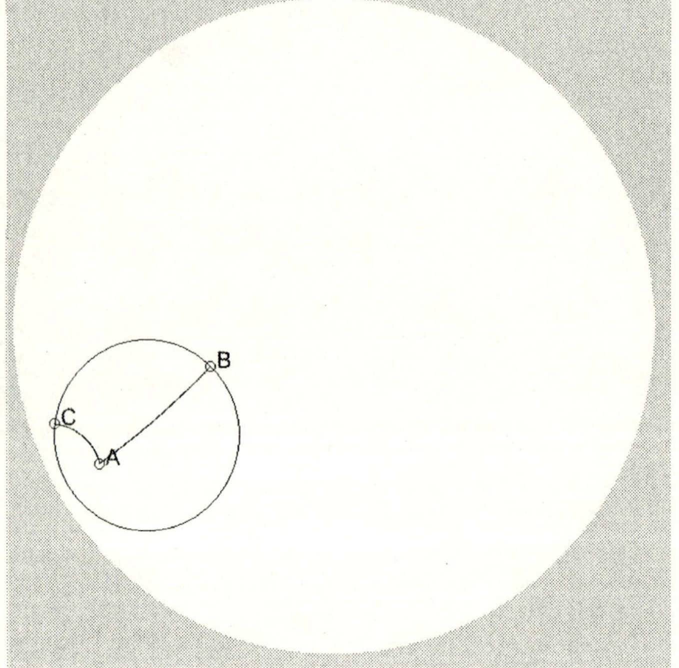

Figure 1-8 shows two "lines" in this model of hyperbolic space. The universe is the interior of the circle C_2. We see "line" $\overleftrightarrow{PQ}$ as a circular arc with endpoints A and B that's orthogonal to C_2, and "line" $\overleftrightarrow{FG}$ a diameter with endpoints E and H, but that's because we're looking with distorted Euclidean vision. To a hyperbolic Flatlander, both "lines" would look like genuine straight lines that are infinite in length. The Flatlander would not be aware of points A, B,

26

E, or H at all, and would, if pressed, admit to them only as mythical "points at infinity."

In Figure 1-8, point A isn't technically in Poincaré Disk space, since it is not *inside* the disk. It's *on* the disk. For the line $\overrightarrow{PQ}$, A is called the *point at infinity* for point P. Similarly, B is called the *point at infinity* for point Q.

How do we measure the distance between two points? Given two points, X and Y, the distance between these points is given by the formula:

$$d(X,Y) = \ln\left[\left(\frac{XY'}{YY'}\right)\left(\frac{YX'}{XX'}\right)\right]$$

where X' and Y' are the points of infinity for X and Y, respectively, and XX', YY', etc., represent the *Euclidean* distances.

Figure 1-7: Congruent bats (or congruent angels) in M.C. Escher's "Circle Limit IV."

Universe 1: Neutral Euclid

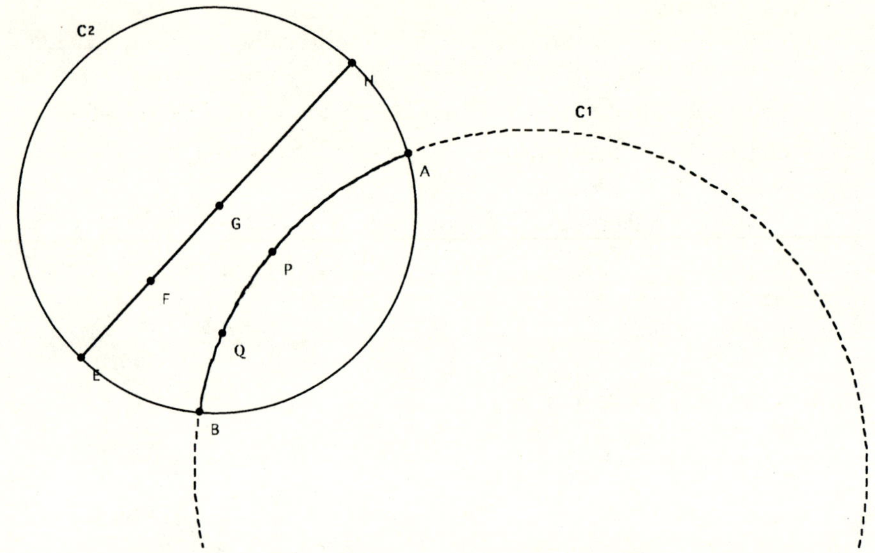

Find the length of segment $\overline{PQ}$ in Figure 1-8 in terms of P, Q, A, and B.

Task 8: It turns out that Poincaré Disk Space is a model for Einstein's Special Theory of Relativity. But to understand this, we need to know something about hyperbolic trigonometric functions. So, here we go!

Here are the definitions of cosh (the hyperbolic cosine), sinh (the hyperbolic sine) and tanh (the hyperbolic tangent.) Note: "cosh" rhymes with "gosh," "sinh" sounds like "cinch," and "tanh" rhymes with "ranch."

$$\cosh(\alpha) = \frac{e^{\alpha} + e^{-\alpha}}{2}$$

$$\sinh(\alpha) = \frac{e^{\alpha} - e^{-\alpha}}{2}$$

$$\tanh(\alpha) = \frac{\sinh(\alpha)}{\cosh(\alpha)}$$

(a) If you know calculus, find the derivatives of the sinh and cosh functions. Kinda' interesting, huh?

(b) Everyone can do this one. Find $\cosh^2(\alpha) - \sinh^2(\alpha)$.
Simplify. Simplify a lot. Kinda' interesting, huh? What does it remind you of?

(c) Why do you suppose these called *hyperbolic* trig functions?

(d) If you know calculus (especially Taylor Series), show that $\cosh(\alpha) = \cos(i\alpha)$, and $\sinh(\alpha) = -i\sin(i\alpha)$.

(e) Prove the following hyperbolic trig identities:

(i) $\sinh(-\alpha) = -\sinh(\alpha)$

(ii) $\cosh(-\alpha) = \cosh(\alpha)$

(iii) $\sinh(\alpha + \beta) = \sinh\alpha\cosh\beta + \cosh\alpha\sinh\beta$

(iv) $\cosh(\alpha + \beta) = \cosh\alpha\cosh\beta + \sinh\alpha\sinh\beta$

(v) $\tanh(\alpha + \beta) = \dfrac{\tanh\alpha + \tanh\beta}{1 + \tanh\alpha\tanh\beta}$

(vi) $\tanh(\alpha - \beta) = \dfrac{\tanh\alpha - \tanh\beta}{1 - \tanh\alpha\tanh\beta}$

Universe 1: Neutral Euclid

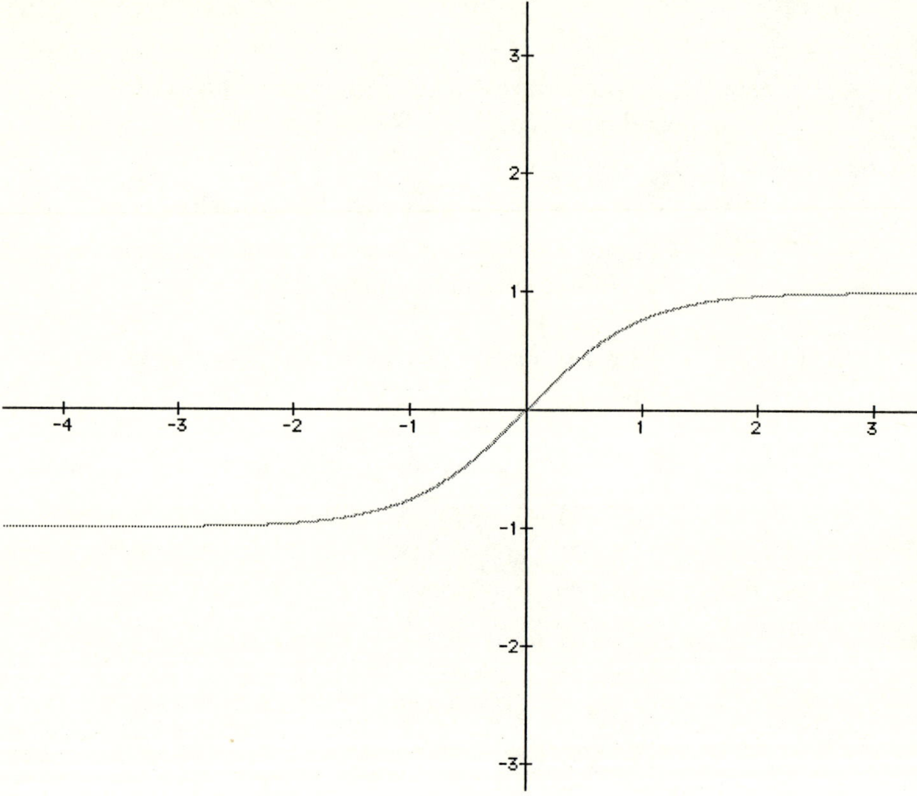

Task 9: By using calculus, or by just looking at the graph in Figure 1-9, one can see that $f(x) = \tanh(x)$ is a strictly increasing function for all x. Therefore, an inverse function exists. We call this function the arctanh(x), or $\tanh^{-1}(x)$.

(a)   By looking at Figure 1-9, sketch the graph of the arctanh(x).

(b)   What is the domain of the arctanh function?

(c)   Show that $\operatorname{arctanh}(v) = \dfrac{1}{2}\ln\left(\dfrac{1+v}{1-v}\right)$.

Notational note: Some British books call this function *artanh* instead of *arctan*. (They leave out the "c".) Similarly, they say *arsin* instead of *arcsin* and *arcos* instead of *arccos*. They also write *colour* instead of *color*, *enrol* instead of *enroll*, they eat pies for dinner, and they eat baked beans for tea.

Figure 1-10: Polar coordinates in Poincaré Disk Space

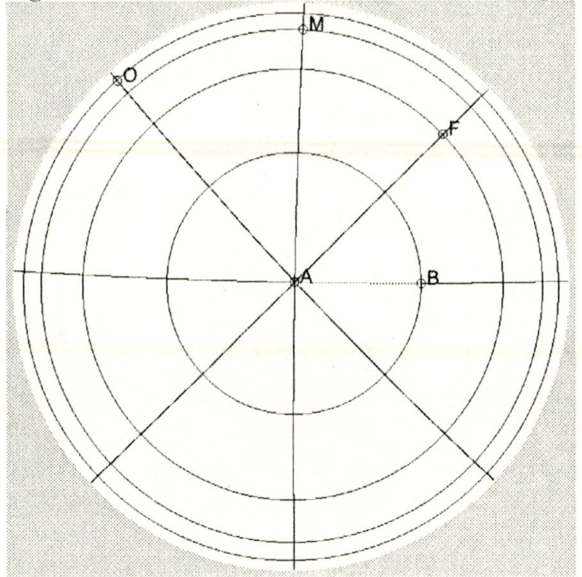

Task 10: One common way to describe how fast an object is moving is to give its speed. Thus, we can say that a car is moving at 65 miles per hour. *Miles* and *hours* are the units, or the dimensions. (The words "unit" and "dimension" are unfortunate in this context, because both of those words already have so many other meanings. Oh well.) What we seek is a unitless way to describe how fast an object is moving.

One way is to measure the speed of an object as a fraction of the speed of light, which is the fastest speed in our universe. So we can say that an object moves at a speed of 0.46, which means that it moves at 46% of the speed of light. (Light moves at about 1 billion miles per hour, or 300, 000 meters per second.)

Another unitless way of describing how fast an object is moving is to give its *rapidity*. The rapidity α of an object is defined by the equation:

$$\alpha = \operatorname{arctanh}(v)$$

where v is the speed of the object given as a fraction of the speed of light. So if an object is moving at a speed of 0.46, or 46% of the speed of light, its rapidity is $\alpha = \operatorname{arctanh}\left(0.46\right) \approx 0.49731$.

We can set up a polar coordinate grid on the Poincaré Disk. We can imagine that each point on the disk represents a velocity on the plane. For example, the point $(2, 45°)$ would represent an object going with rapidity 2 heading 45° "north" from the polar axis. This point is represented by point D in Figure 1-10. So the Poincaré distance between a point and the origin A we interpret as the rapidity of an object, and the polar angle represents the direction of the motion on the plane.

(a)   Faramir is piloting a spaceship in the outer space of Flatland. He is traveling at a velocity $(2, 45°)$, which is represented by point F in Figure 1-10. At what fraction of the speed of light is Faramir traveling, from our point of view?

(b)   Zlinda is piloting a spaceship that's going about 76.159% of the speed of light. Which point in Figure 1-10 represents Zlinda? And what direction is she going?

(c)   Melinda is represented by point M, and Olivia is represented by point O. What percentage of the speed of light is Melinda traveling? Olivia?

(d)   Are there going to be any points in Poincaré Disk Space that will violate the postulate from relativity that says that nothing can travel faster than the speed of light? Explain your answer.

Task 11: Every statement in Einstein's Special Theory of Relativity can be translated into an equivalent statement about the Poincaré Disk, and vice-versa. (In math lingo, Special Relativity and the Poincaré Disk are *isomorphic*.) To make the translation, just remember that "points" are particles in uniform motion, which can be represented by rapidity velocities in polar coordinates, while "lines" are the "lines" of the Poincaré Disk. So, without knowing it, you've already been studying relativity!

                              Universe 1: Neutral Euclid

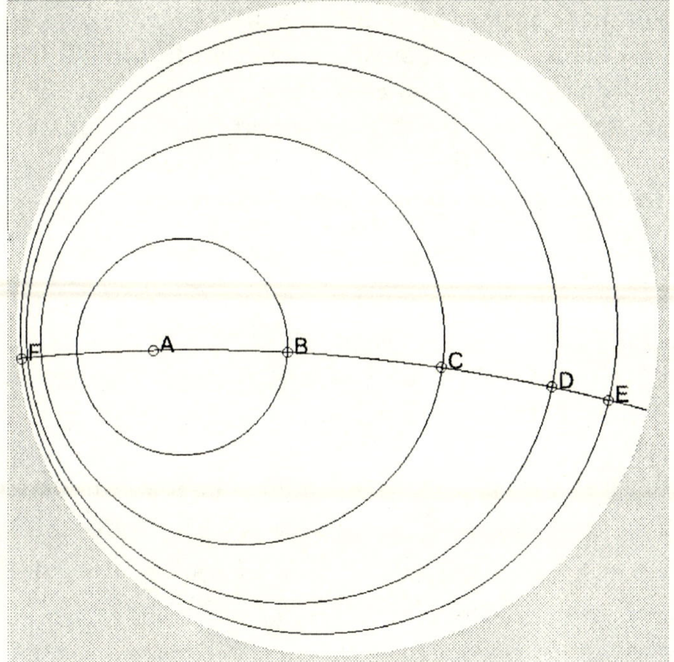

In Figure 1-11, all of the labeled points are along the polar axis with origin A.

(a)   What are the coordinates of each point?

(b)   What percentage of the speed of light does A think that B is going?

(c) B, of course, is free to think that she's the one who's standing still and A's the one who's moving. In other words, she can think that she's the one who's at the origin, and draw her own polar grid. What percentage of the speed of light does B think that A is going? And in what direction?

(d)   How fast (as a percentage of the speed of light) does A think that C is going?

(e)   How fast (as a percentage of the speed of light) does B think that C is going?

(f)   You have an imaginary friend named Harvey, who is reading these problems with you. Harvey is slow, but he's cute. He says, "I don't understand your answer to part (e). Just take the answer to part (d), and subtract the answer to part (b). So, B thinks that C is going about 20.3% of the speed of light." You think about this, and realize that Harvey has a point. What's wrong with Harvey's reasoning?

(g)   From A's point of view, E is traveling at 99.9% of the speed of light to the right, and F is traveling at 99.9% of the speed of light to the left. So does E think that F is traveling *faster* than light? How fast does E think that F is traveling?

(h)   Suppose that you are traveling in your spaceship in a straight line at a speed (from my point of view) of $v_1 = \tanh \alpha$, and you fire a torpedo at a meteor directly in front of you. You fire the torpedo at a speed (from your point of view) of $v_2 = \tanh \beta$. Let v_3 be the speed of the torpedo from my point of view. You probably realize by now that, despite common sense, $v_3 \neq v_1 + v_2$. (Toto, I think we're not in Euclidean Space anymore...) Show that

$$v_3 = \frac{v_1 + v_2}{1 + v_1 v_2}$$

[Hint: use the addition formula for tanh.] This is the famous Einstein Addition Law.

(i)   Suppose I think that you are traveling in a straight line at 90% of the speed of light, and that *you* think that you've just fired a torpedo straight ahead that is going at 80% of the speed of light. Use the Einstein Addition Law to find out how fast I think that the torpedo is going.

(j)   In this book, we are taking the convention that all speeds v are given in terms of a fraction of the speed of light. If you want to measure speeds using units (such as meters per second) the formula in part (h) will need to be altered. Let c be the speed of light,

                                          Universe 1: Neutral Euclid

measured in whatever units you wish to use. Show that the general formula of the Einstein Addition Law becomes:

$$v_3 = \frac{v_1 + v_2}{1 + \left(v_1 v_2 / c^2\right)}$$

(k)   Your friend Harvey says: "This is ridiculous! If I'm on a train going 50 miles per hour, and I throw a ball towards the front of the train at 20 miles per hour, then obviously the ball is going at 50 + 20 = 70 miles per hour. Einstein's formula is bunk." How would you respond to Harvey?

Figure 1-12: Construction of equilateral triangles. The two circles are congruent, and the two triangles are congruent. (The two triangles are also equilateral.)

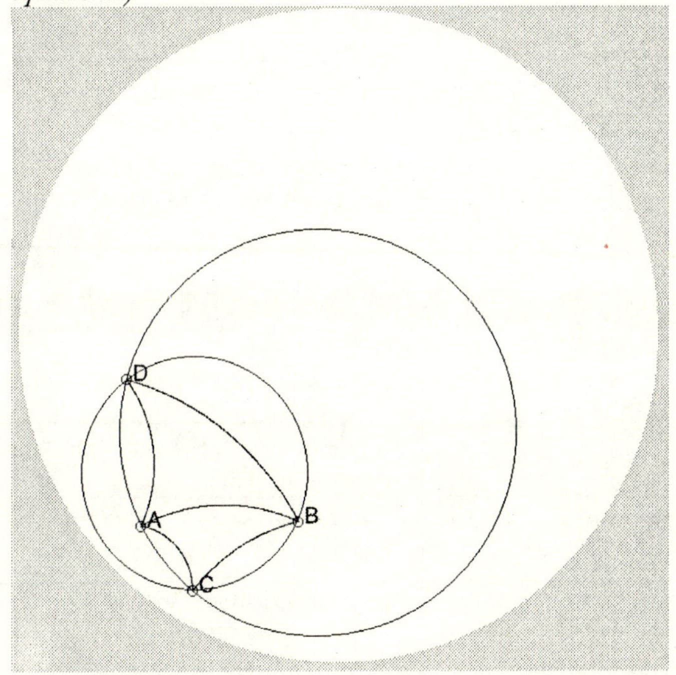

(l)   Harvey is still not quite satisfied by your answer to part k. He says, "When you fired the torpedo, you must have made a mistake in measuring its speed. Maybe something happens to clocks in this universe, as you approach the speed of light. After all, if your

Universe 1: Neutral Euclid

clock was running—." There's a sudden crash of thunder. "—that would explain everything." Unfortunately, you missed hearing one of Harvey's words, because of the crash of thunder. What word do you think it was?

Task 12: (This is good preparation for Task 13.) Given a segment $\overline{AB}$, construct a point C with a straightedge and compass such that $\triangle ABC$ is equilateral. [After you finish this problem, you can check your answer, sort of, by looking at Figure 1-12. It's a picture of the construction done by a creature in Poincaré Disk Space. If you're wearing your Poincaré glasses (not included with the purchase of this book) you'll notice that both of the "circles" are "congruent", as are both of the "triangles" $\triangle ABC$ and $\triangle ABD$.]

Task 13: Here is a modern rewording of Euclid's very first proposition:

Proposition 1: Given a segment $\overline{AB}$, there exists a point C such that triangle $\triangle ABC$ is equilateral.

The following proof is due to Euclid. It is put in two-column form, with modern wording. Fill in the blanks. There is one blank that you will not be able to fill in. Put "uh-oh!" in that blank.

Statements	Reasons
(1) There exists a circle α with center A and radius $\overline{AB}$.	(1) --?--
(2) --?--	(2) Postulate 3 (Or Set Theory. See note to Task 6.)
(3) There exists a point C where α and β intersect.	(3) --?--
(4) Since C is on α, --?--. Since C is on β, --?--.	(4) Definition of "circle"
(5) $\triangle ABC$ is an equilateral triangle	(5) --?--

Task 14: The first four of Euclid's postulates do not suffice. It's clear that we need another postulate in order to prove Euclid's Proposition 1. Write a postulate that will serve.

     Universe 1: Neutral Euclid

Task 15: Interpret Proposition 1 in terms of special relativity. Don't use the words *segment, point, triangle, equilateral,* or *congruent.*

Task 16: In Figure 1-13, A is the center of both circles. The smaller circle has radius 1, and the larger circle has radius 2.

Suppose A, B, and C are all at the same point when time $t = 0$. A thinks she is standing still, and that B and C are moving to the right.

(a)   One hour later (from A's point of view) how far away is B from A? (Light travels at 1 billion miles per hour.) Give your answer in millions of miles.

(b)   One hour later (from A's point of view) how far away is C from A?

(c)   One hour later (from A's point of view) what is the distance between B and C?

(d) B thinks that he is standing still, that A is moving to the left, and that C is moving to the right. One hour later (from B's point of view) how far away from B is A?

(e)   One hour later (from B's point of view) how far away from B is C?

(f)   One hour later (from B's point of view) how far away from C is A?

(g)   Harvey (your imaginary friend) says, "Wait a second! Shouldn't part (c) and (e) have the same answer? And shouldn't part (b) and (f) have the same answer? What's going on here?" How would you answer Harvey?

Figure 1-13: Harvey doesn't understand why distances are relative.

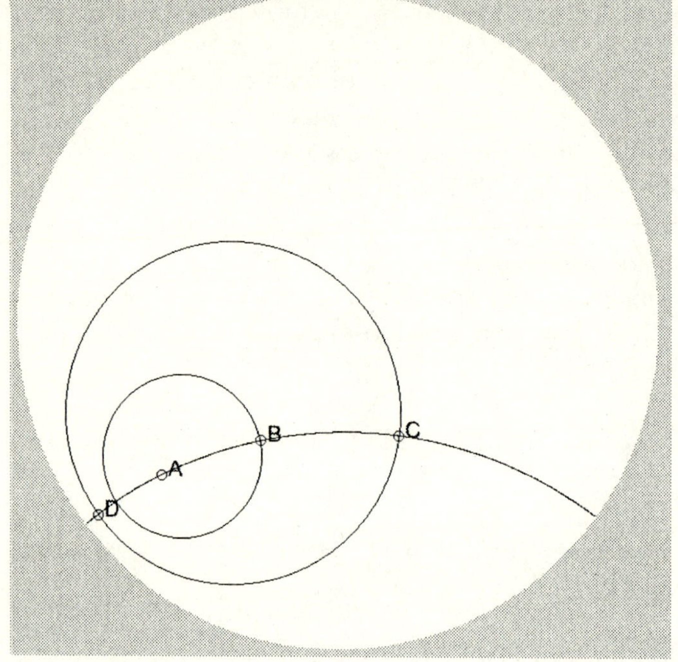
B
A
C
D

Universe 2: Neutral With Continuity

In Which We Prove Euclid's First Three Propositions, and Make a Feeble Attempt to Prove His Fourth

The value of Euclid's work as a masterpiece of logic has been very grossly exaggerated.
--Bertrand Russell

Three of Euclid's Postulates

1. For every point P and for every point Q not equal to P there exists a unique line l that lies on P and Q.

2. Given segments $\overline{AB}$ and $\overline{CD}$, there exists a unique point E such that B is between A and E (or $A*B*E$), and $\overline{CD} \cong \overline{BE}$.

4. All right angles are congruent to each other.

Circular Continuity Axioms:

(1) If a circle α with center A has one point inside and one point outside another circle β with center B, then the two circles intersect in exactly two points, and these two points are on opposite sides of $\overline{AB}$.

(2) If circle α with center A and another distinct circle β with center B intersect at two points, then they intersect at exactly two points, and these two points are on opposite sides of $\overline{AB}$.

Task 1: Notice that we've omitted Euclid's Postulate 3. We don't need it. (See Task 6 in Universe 1.) Perhaps we don't need to take the word *line* as an undefined term, for similar reasons. We defined the word *circle* using set theory. Maybe we can define the word *line* using set theory as well.

(a)   If we could get rid of the word *line* as an undefined term, it would be a good thing. Why?

(b)   Let's try defining the word *line*. Let's define the line $\overleftrightarrow{AB}$ by the set of points in the segment $\overline{AB}$, along with the set of points C such that $A*B*C$, and the set of points D such that $D*A*B$. (Notice that the *segment* $\overline{AB}$ was previously defined using the word *between* and set theory.) Does this work? Or is there something wrong?

Note: The word *between* is funny, and the English language does not deal with it gracefully. When we say "B is between A and C," B *isn't* between in the way we *say* it; we say "B" first, which is very confusing, and always causes a mental hiccup. Better to say it mathematically, $A*B*C$, which says the same thing, but can be read "A then B then C." By the way, $A*B*C$ and $C*B*A$ mean the same thing.

Task 2:   The definitions in this task will be referred to again and again. Give them a prominent place in your glossary. Put a gold star next to them, or something.

(a)   First we will define what we mean when we say that one segment is *longer* than another segment. In your definition, do not mention any kind of *measurement*. That's what part (b) is about. Use your undefined terms. I'll start you off: "We say that $\overline{AB} > \overline{CD}$ if...

(b)   *Now* we will define what we mean by the *measure of segment* $\overline{AB}$, denoted by $m(\overline{AB})$, or simply by AB. Obviously, there are lots of schemes that one can work out for measuring the length of segments. One can measure in feet, inches, meters, light years, or furlongs, for example. But whatever scheme you work out, it needs to fulfill the following three properties.

The *measure of segment* $\overline{AB}$ is a positive real number such that:

(i) $\overline{AB} \cong \overline{CD}$ iff $AB = CD$

(ii)   point G is between A and B iff...

(iii) $\overline{AB} > \overline{CD}$ iff $AB > CD$

Your task: finish sentence (ii). Hint: Think segment addition.

Task 3:   Define what it means to be "inside" and "outside" of a circle. I'll start you off. "Let α be the circle with center A and radius AB. The point E is *inside* the circle α if..."

Task 4: Now we are finally ready to prove Euclid's very first proposition. Fill in the blanks in the two-column proof below.

Proposition 1: Given a segment $\overline{AB}$, there exists a point C such   that triangle $\triangle ABC$ is equilateral.

Statement	Reason
(a) There exists circle α with center A and radius AB, and circle β with center B and radius BA.	(a) --?--
(b) There exists a point E such that --?-- and $BA = AE$.	(b) Postulate 2
(c) $EB > BA$	(c) --?--
(d) --?--	(d) Definition of "outside"
(e) E is on circle α	(e) --?--
(f) --?--	(f) Definition of "inside"
(g) Point B is on circle α	(g) --?--
(h) --?--. Call one of them C.	(h) Circular Continuity Axiom
(i) Since C is on α, $AB = AC$. Since C is on β, $BA = BC$.	(i) --?--
(j) --?--	(j) Definition of equilateral triangle

Task 5:   This problem refers to Figure 2-1. A and B are the centers of their respective circles. Suppose A thinks that B is

traveling at a velocity of (1, 0°). [Note: in this book, the first component of velocity vectors will always be the rapidity.]

(a)	How fast, as a percentage of the speed of light, does A think that B is going?

(b)	How fast, as a percentage of the speed of light, does B think that A is going?

(c)	How fast, as a percentage of the speed of light, does A think that C is going?

(d)	How fast, as a percentage of the speed of light, does C think that B is going?

(e)	The measure of $\angle CAB$ is about 52.6°. From A's point of view, what is the velocity of C?

(f)	Restate Proposition 1 in terms of special relativity. Don't use the words *segment, point, triangle, equilateral,* or *congruent.* I'll start you off: "Given two velocities A and B..."

(g)	Every theorem in Neutral Geometry is true in both Euclidean Geometry and Special Relativity. Why?

Task 6:	We will now prove Euclid's second proposition. Fill in the blanks in the two-column proof below.

Proposition 2: Let $\overline{BC}$ be a given segment, and let A be a given	point not on $\overline{BC}$. Then there exists a point D such that $BC = AD$.

Proof:

Statement	Reason
(1) There exists a segment $\overline{AB}$.	(1) --?--
(2) --?--	(2) Postulate 2

Note: If you look at Euclid's original proof of Proposition 2, you might be surprised to see that it takes up an entire page. Why is our proof so much shorter and simpler? This is our payoff for using a more rigorous Postulate 2. Euclid's original Postulate 2 simple said that we can "produce a finite straight line continuously in a straight line," which is a rather vague and imprecise statement. Thus he wastes a lot of space in his proof of Proposition 2 in trying to make Postulate 2 clear.

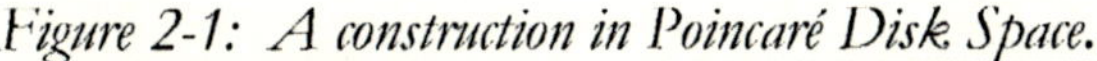

Figure 2-1: A construction in Poincaré Disk Space.

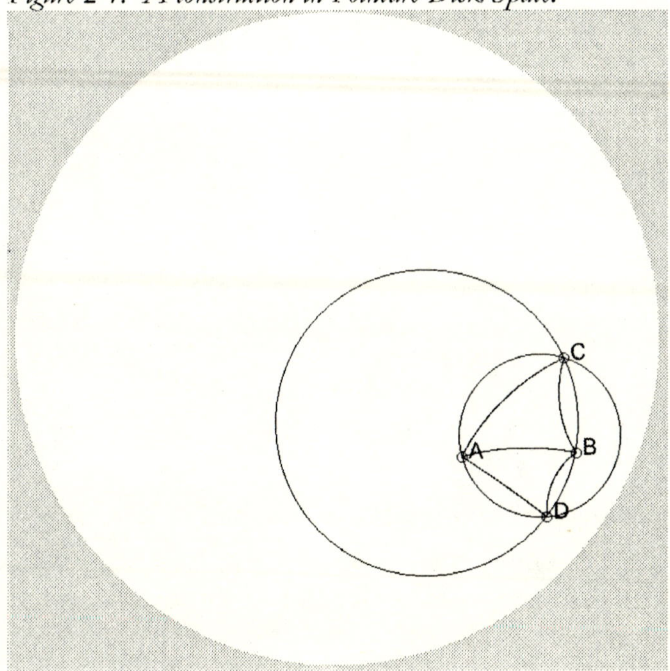

Task 7: Is Proposition 2 true in Universe 1?

Task 8: Refer to Figure 2-2.

(a) From A's point of view, B is traveling at a velocity of (0.75, 4°). What is the relative speed between A and B? (In other words, how fast, as a percentage of the speed of light, does each one think that the other is going?)

(b) Does there exist a velocity D such that the relative speed between C and D is the same as your answer to part (a)? Justify your answer.

Universe 2: Neutral With Continuity

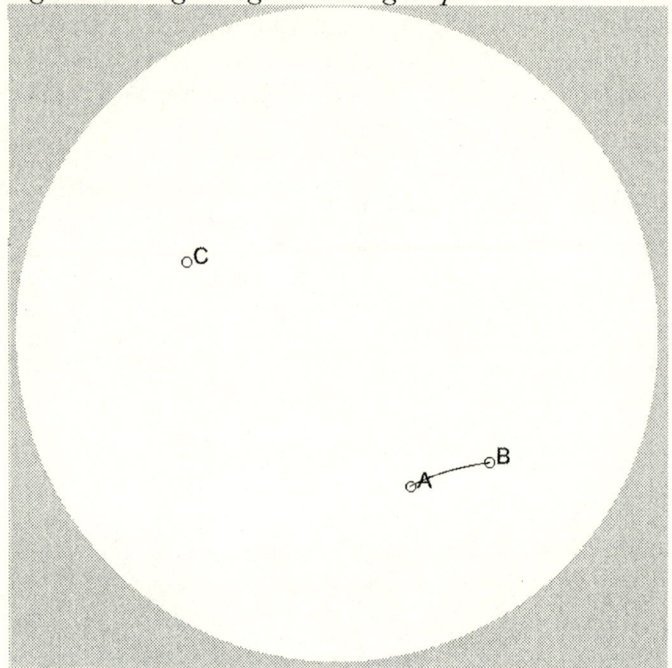

Task 9: Here is a modern rewording of Euclid's third proposition:

> Proposition 3: Given two segments: $\overline{AB}$ and $\overline{CG}$, with $AB > CG$. Then there exists a point E between A and B such that $AE = CG$.

This proposition can be proved in one step. Do so.

Note: You may wonder why I chose the letters A, B, C, and G instead of A, B, C, and D. Answer: I didn't choose those letters; Euclid did. Why? I don't know.

Another Note: If you happen to have a copy of *The Elements* handy, you'll again notice that our proof of Proposition 3 is so much shorter and simpler than Euclid's original. This is more payoff for our extra rigor.

Task 10: Refer to Figure 2-3.

(a)   True or false: From looking at the picture, it's clear that $CD > AB$.

(b) A thinks that B is going at a velocity of $(1, -60°)$. What's the relative speed between A and B?

(c) C thinks D is going at a velocity of $(0.75, 0°)$. What's the relative speed between C and D?

(d)   Does there exist a velocity E between A and B such that the relative speed between A and E equals the relative speed between C and D? Justify your answer.

(e)   What's the velocity E, from A's point of view?

(f) $AE = ?$

Figure 2-3: Comparing relative speeds.

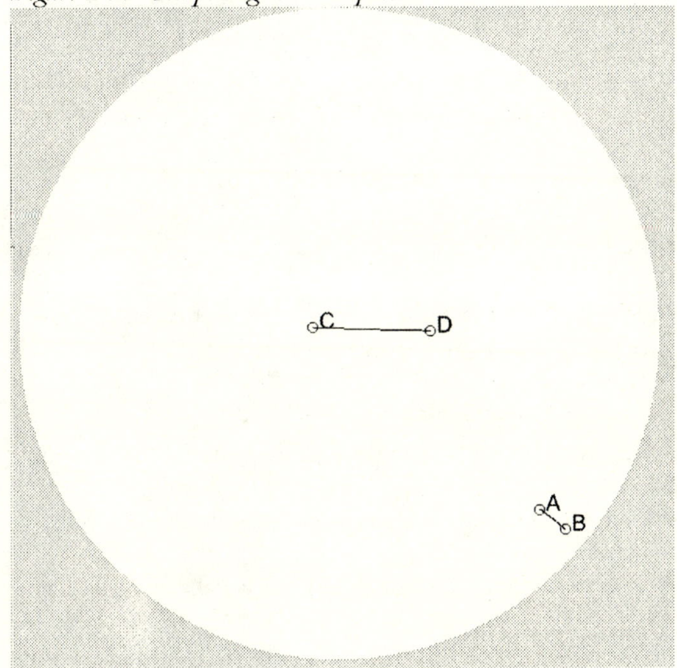

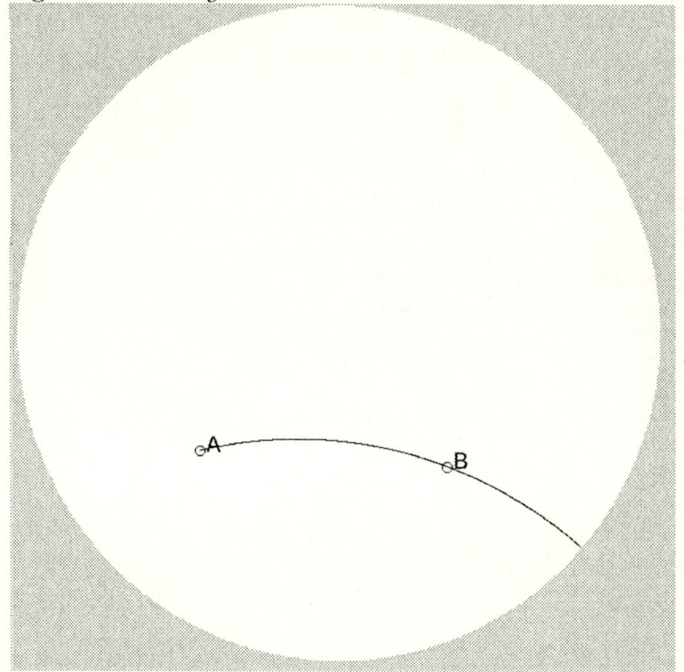

Task 11: Define *ray*. I'll start you off. "The *ray* $\overrightarrow{AB}$ is the following set of points on the line $\overleftrightarrow{AB}$: those points that..." [Note: A is called the *vertex* of ray $\overrightarrow{AB}$.] If it will help, Figure 2-4 is the picture of a "ray," according to the creatures in Poincaré Disk Space.

Task 12: Define *opposite rays*. If it will help, in Figure 2-5 $\overrightarrow{AB}$ and $\overrightarrow{AC}$ are opposite rays (according to the creatures in Poincaré Disk Space). I'll start you off. Rays $\overrightarrow{AB}$ and $\overrightarrow{AC}$ are called *opposite rays* if...

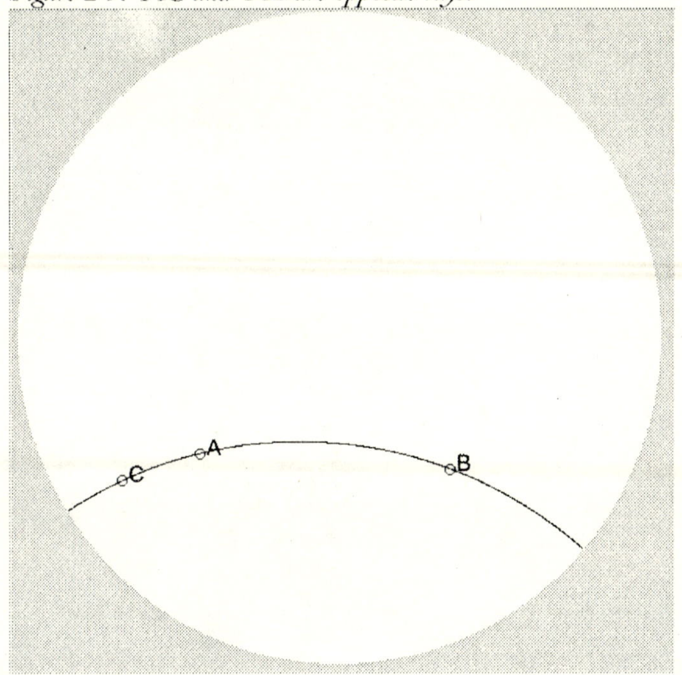

Task 13: Postulate 4 uses the word *angle*. Complete the following definition of *angle*. While you're at it, also define what you mean be the *vertex* and the *sides* of an angle. I'll start you off. "The *angle* $\angle CAB$ is...." There's a "helpful" picture of angle $\angle CAB$ in Figure 2-6.

Note: In Euclid's *Elements*—and in this book—there is no such thing as a "straight angle," an angle whose measure is π radians or 180°. All angles in *The Elements* (and in this book) will have a measure of less than 180°.

Another Note: The angle $\angle CAB$ is the same as the angle $\angle BAC$.

Figure 2-6: The angle $\angle CAB$.

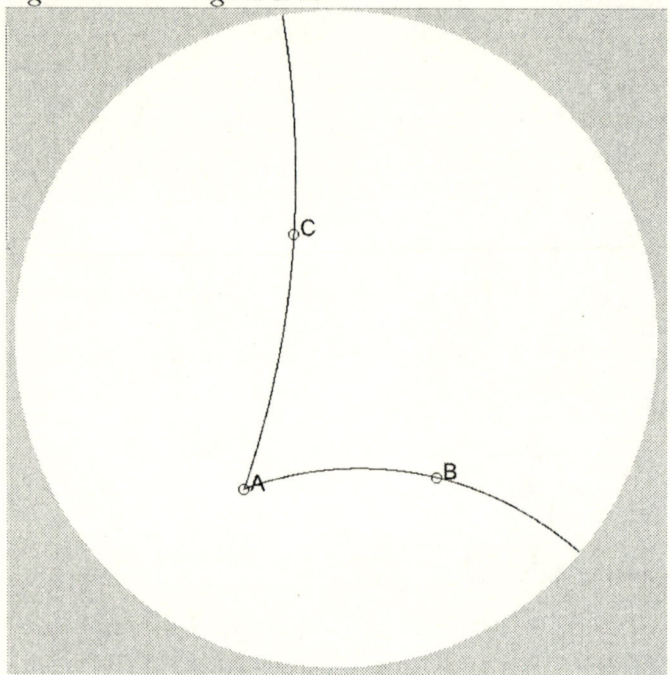

Task 14: We don't define what we mean by *congruent segments* or *congruent angles*. These are undefined terms. But if we talk about other things being congruent, we need to define what we mean. Define *congruent triangles*. I'll start you off. "Given two triangles, $\triangle ABC$ and $\triangle DEF$. We say these two triangles are *congruent*, and we write $\triangle ABC \cong \triangle DEF$ if...."

Task 15: In Figure 2-7, the two triangles are congruent. So $\triangle ABC \cong \triangle\,-?-$. Replace the questions mark with the appropriate letters. Note: the order of the letters *does* make a difference! Corresponding parts must be congruent. For example, *EDF* is a wrong answer.

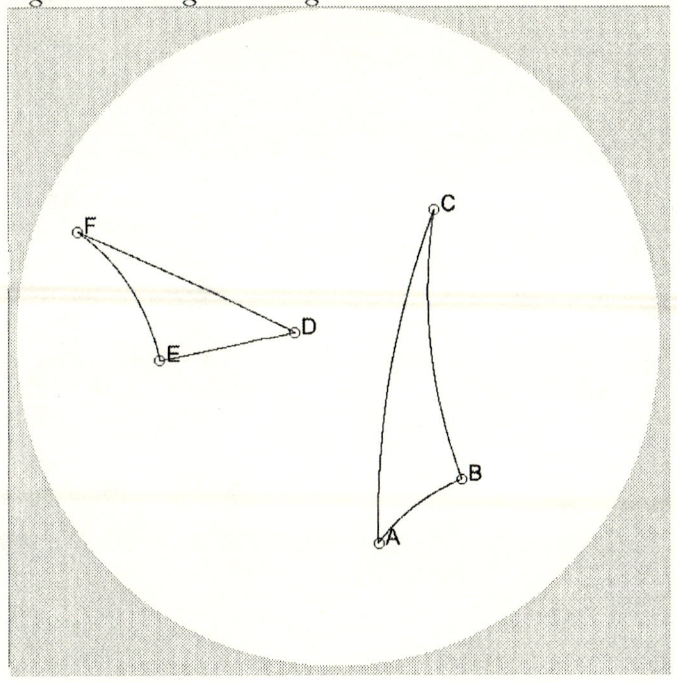

Task 16: Define *congruent circles*. [Yes, yes, the two circles in Figure 2-8 are "congruent." And points A and C are in their respective "centers."]

Task 17: Here is a modern rewording of Euclid's fourth proposition:

> Proposition 4 (SAS Theorem): Given two triangles $\triangle ABC$ and $\triangle DEF$. If $AB = DE$ and $AC = DF$, and if $\angle A \cong \angle D$, then the two triangles are congruent.

Try proving it yourself, using the postulates of Universe 2, as well as the three previously proved propositions. But don't try to hard, for SAS is unprovable in this universe. Indeed, in order to prove SAS there seems to be some additional postulate that we need. Think of one that will serve.

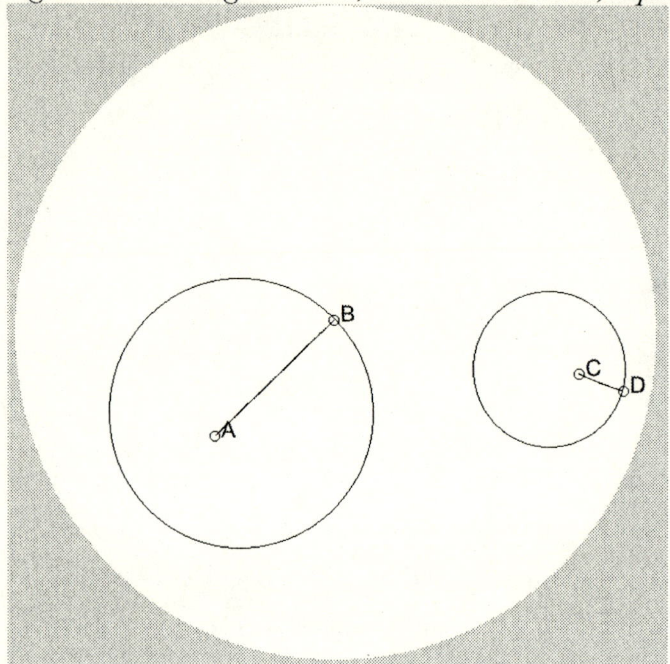

Note: If you happen to have a copy of *The Elements* handy, you might want to read Euclid's embarrassingly bad "proof" of Proposition 4. When I first read it, I threw the book across the room in disgust. However, don't write Euclid off; his proofs get much better—even eloquent—as the book progresses.

Universe 3: Neutral With SAS

In Which We Take SAS as a Postulate, and, After Proving a Few More of Euclid's Propositions, Make a Valiant Attempt to Prove SSS

We have Einstein's space, de Sitter's space, expanding universes, contracting universes, vibrating universes, mysterious universes. Indeed, the pure mathematician can create universes just by writing down an equation, and if he is an individualist he can have a universe of his own.

--J. J. Thomson (English physicist)

Three of Euclid's Postulates

1.   For every point P and for every point Q not equal to P there exists a unique line l that lies on P and Q.

2.   Given segments $\overline{AB}$ and $\overline{CD}$, there exists a unique point E such that B is between A and E (or $A * B * E$), and $\overline{CD} \cong \overline{BE}$.

4.   All right angles are congruent to each other.

Circular Continuity Axioms:

(1)   If a circle α with center A has one point inside and one point outside another circle β with center B, then the two circles intersect in exactly two points, and these two points are on opposite sides of $\overline{AB}$.

(2)   If circle α with center A and another circle β with center B intersect at two points, then they intersect at exactly two points, and these two points are on opposite sides of $\overline{AB}$.

SAS Axiom: If two sides and the included angle of one triangle are congruent respectively to two sides and the included angle of another triangle, then the two triangles are congruent.

Figure 3-1: Congruent triangles

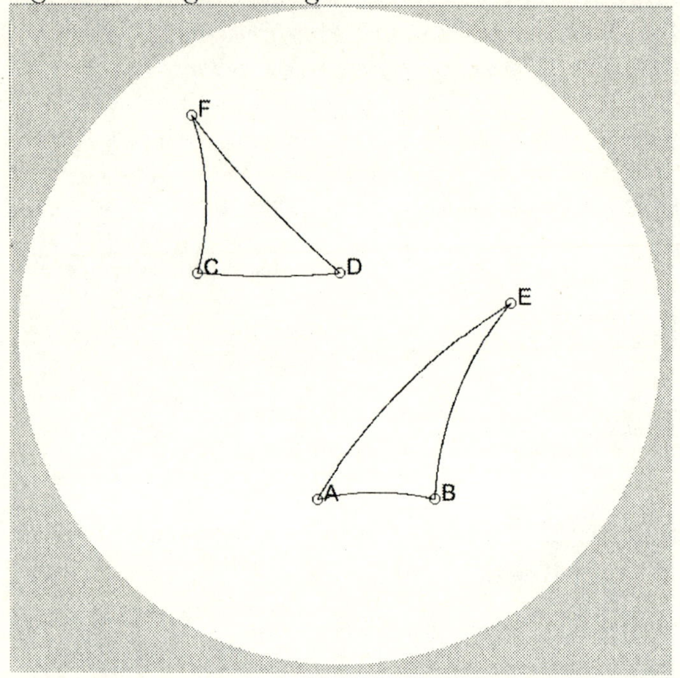

Task 1: Refer to Figure 3-1. A thinks B is going at a velocity of (1, 0°), and E is going at a velocity of (2, 45°). D thinks C is going at a velocity of (1, 180°) and F is going at a velocity of (2, 135°).

(a) $\triangle ABE \cong \triangle$--?--, by --?--.

(b) B thinks that E is traveling at 93.4% of the speed of light. Name another pair of velocities that have this same relative speed. Justify your answer.

Task 2: We are now ready to prove Euclid's fifth proposition. Fill in the blanks in the two-column proof below. Note: The proof below is due to Pappus, another ancient scholar. It is superior to Euclid's proof.

> Proposition 5 (The Isosceles Triangle Theorem):   In an isosceles triangle, the base angles are congruent.

Statement	Reason
(1) Let $\triangle ABC$ be an isosceles triangle, with $AB = AC$.	(1) --?--
(2) --?--	(2) SAS Axiom
(3) $\angle B \cong \angle C$	(3) --?--

Figure 3-2: Illustration for Task 3

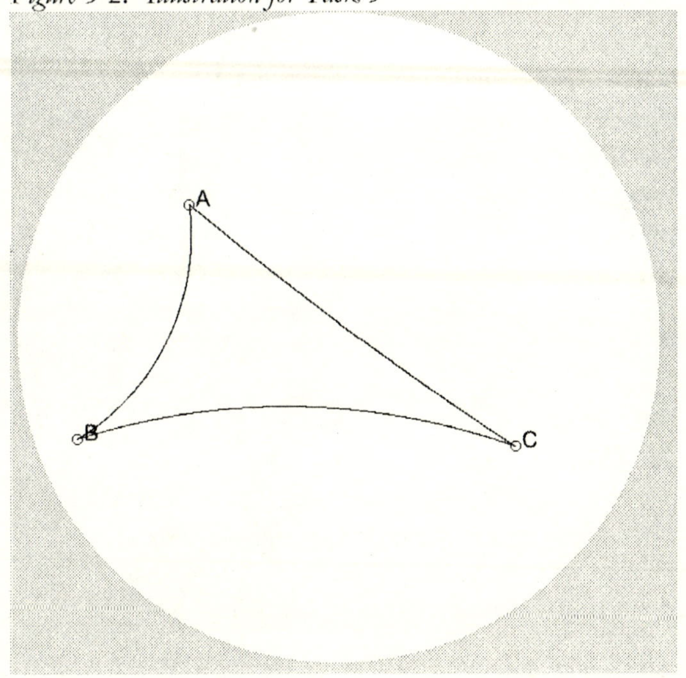

Task 3: Refer to Figure 3-2. B thinks C is traveling at a velocity of $(4.038, 0°)$ and A is traveling at a velocity of $(3, 14°)$. C thinks B is traveling at a velocity of $(4.038, 180°)$. Moreover, C thinks that A is traveling at 99.50547537% of the speed of light. What velocity does C think that A is traveling? Justify your answer.

Task 4: Some questions to ponder...

(a)   Is Proposition 5 valid in Universe 1? In Universe 2? If not, what additional axiom(s) do you need?

(b)   In Universe 3, is an equilateral triangle necessarily equiangular? Why, or why not?

Task 5: Complete the definition of *supplementary angles*. "If two angles $\angle BAD$ and $\angle CAD$ have a common side, ray $\overrightarrow{AD}$, and the other two sides, rays $\overrightarrow{AB}$ and $\overrightarrow{AC}$ are..."

Note: In elementary geometry, two angles are called "supplementary" if their angle measures add up to π, or 180°. See if you can define *supplementary* without referring to the terms "radians" or "degrees" (or any other kind of angle measure). Euclid never talks about radians or degrees. Hint: Use the word *congruent* in your definition.

Task 6: Define *right angle*. I'll start you off. "An angle is called a *right angle* if..." (Again, don't talk about the angle measure equaling $\dfrac{\pi}{2}$, or 90°. Don't talk about *angle measure* at all!)

Task 7:   Given that your definition of right angle in Task 6 doesn't refer to degrees or radians, do we even know that all right angles are congruent to each other?

Task 8: This task will be referred to again and again.

(a)   First we will define what we mean when we say that one angle is *bigger* than another angle. In your definition, do not mention any kind of *measurement*. That's what part (b) is about. Use your undefined terms. I'll start you off: "We say that $\angle ABC > \angle DEF$ if..."

(b)   Now we will define what we mean by the *measure of angle* $\angle RST$, denoted $m\angle RST$. Obviously, there are lots of schemes that have already been worked out for measuring angles. One can measure angles in degrees, radians, or gradients for example, and Euclid measured all of his angles in terms of right angles. ("The sum of the angles in any triangle is equal to two right angles," says he.)

54

But whatever scheme you work out, it needs to fulfill the following three properties.

The *measure of angle* $\angle RST$ is a positive real number such that:

(i) $\qquad \angle RST \cong \angle UVW$ iff $m\angle RST = m\angle UVW$

(ii) $\qquad m\angle RST < m\angle UVW$ iff $\angle RST < \angle UVW$

(iii) $\qquad$ ray $\overrightarrow{SG}$ is between rays $\overrightarrow{SR}$ and $\overrightarrow{ST}$ iff...

Your task: finish that last sentence.

Task 9: The first part of this task is to actually *read* the following:

Since all right angles are congruent (by Postulate 4) the measures of all right angles are equal, and thus we can assign a unique number to it. For example, we could say that if $\angle ABC$ is a right angle, then $m\angle ABC = 34$ glorphs. Thus, the measure of *all* right angles would be 34 glorphs. As you undoubtedly know, people have already come up with schemes for measuring angles, so we needn't reinvent the wheel here. We can say that $m\angle ABC = 90$ degrees (often denoted 90°), or we can say $m\angle ABC = \dfrac{\pi}{2}$ radians. The word "radians" is usually omitted, since (as we learned in precalculus) radians can be defined as the ratio of the arclength of a sector to its radius. Thus, we can say that $m\angle ABC = \dfrac{\pi}{2}$, where $\dfrac{\pi}{2}$ is a pure number, with no units attached. This again underscores the fact that in Euclidean geometry, there are natural, unit-less ways to measure angles. (Euclid measures all his angles in terms of right angles, for example.) That is why the National Bureau of Standards does not need to keep a platinum right angle as a reference; anyone with a straightedge and compass can construct one for herself.

You may have noticed that in Poincaré Disk Space there is also a natural, unit-less ways of measuring length. This is true in many other non-Euclidean spaces too. So in these universes, we can say that the length of a segment is 5. Not 5 feet or 5 meters, just 5. Hence, in these universes, there would be no need for a National Bureau of Standards at all! We'll say more about this later.

Suppose that $\angle ABE \cong \angle ABD$. Does is necessarily follow that $\angle ABE = \angle ABD$?

Task 10: Here is a modern rewording of Euclid's sixth proposition:

> Proposition 6: (The converse of Prop. 5). If two angles in a triangle are congruent, then the opposite sides are congruent (and thus the triangle is isosceles.)

Fill in the blanks of the following proof. This proof is straight out of Euclid's *Elements*, but—alas!—it contains a fatal flaw. One of the fill-in-the-blank-steps doesn't quite work (although it *is* convincing). Write "uh-oh!" in the blank statement or reason that doesn't work. [Note: "WMAWLOGT" stands for "we may assume without loss of generality that." Also, "RAA" stands for "reductio ad absurdum."]

Statements	Reasons
(a) Consider $\triangle ABC$ with $\angle B \cong \angle C$.	(a) --?--
(b) Suppose --?--. WMAWLOGT $AB > AC$.	(b) RAA hypothesis
(c) There exists a point D between A and B such that $BD = AC$.	(c) --?--
(d) --?--	(d) SAS Postulate
(e) But this is absurd, since $\triangle DBC$ is inside $\triangle ACB$.	(e) --?--
(f) --?--	(g) RAA conclusion

Task 11: Euclid never invokes Proposition 6 in Book I of *The Elements*. (He does invoke it for a proof in Book II.) Hence, there is no logical reason why we must prove it now. We could delay its proof until after Proposition 26 of Book I, which is ASA, the "angle-side-angle" criterion for congruent triangles. Use ASA to prove Proposition 6. This is by far the easiest way to prove it!

Statements	Reasons
(a) Consider $\triangle ABC$ with $\angle B \cong \angle C$.	(a) --?--
(b) --?--	(b) ASA
(c) $AB = AC$	(c) --?--

Figure 3-3: Illustration for Task 12

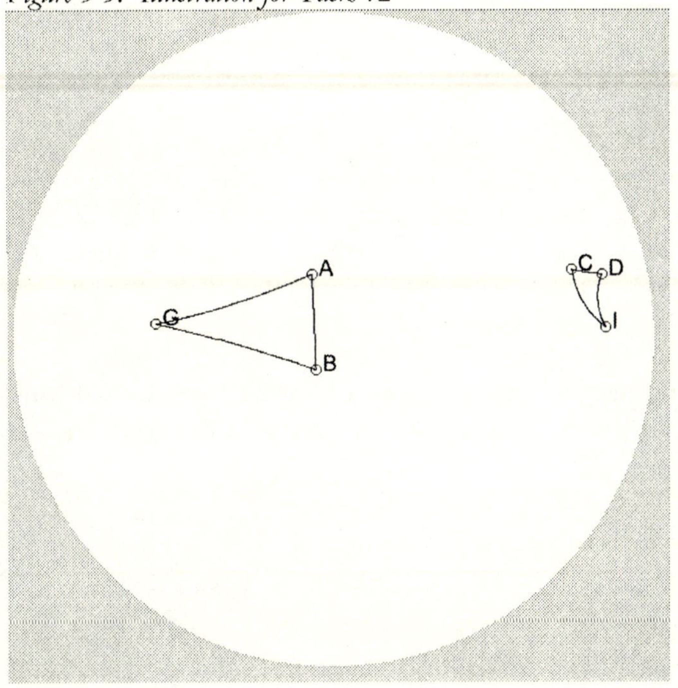

Task 12: Refer to Figure 3-3.

(a) A thinks that G is traveling at a velocity of (1.2, -159.5°), and that B is traveling at a velocity of (0.6, -90°). Moreover, $m\angle B = 69.5°$. At what speed does G think that B is traveling? Justify your answer.

(b) C thinks D is traveling at a velocity of (0.6, 0°), and that I is traveling at a velocity of (1.2, -69.5°). D thinks C is traveling at a velocity of (0.6, 180°). At what velocity does D think that I is traveling?

Task 13: (This is good preparation for Task 14). Using a straightedge, draw an arbitrary angle. Copy the angle (in other words, construct another angle that is congruent to it) using only a straightedge and compass.

Task 14: *Try* to prove the following theorem. You won't be able to, but it's enlightening to give it a go.

> Given $\angle ABC$ and segment $\overline{DE}$. Then there exists a point F such that $\angle ABC \cong \angle DEF$.

While attempting to prove this, try imagining how you would copy an angle using just a straightedge and compass. You still won't be able to prove it, but you *will* notice that there is some theorem or postulate that you need in order to prove it. What theorem or postulate is that?

Task 15: Using just the postulates of Universe 3 (along with our previously proved theorems), try to prove the SSS criterion for congruent triangles:

> Given two triangles $\triangle ABC$ and $\triangle DEF$. *If* $AB = DE$, $BC = EF$, and $AC = DF$, then $\triangle ABC \cong \triangle DEF$.

If you run into difficulties, explain what they are.

Task 16: From the previous two tasks, you're probably getting the idea that Euclid is soon going to attempt to prove SSS. You're right, but we need to do a few things first. Here is Euclid's seventh proposition in modern garb.

> Proposition 7: Let triangle ABC be given. Then there does not exist a point D (where $D \neq C$) on the same side of $\overline{AB}$ as C such that $AC = AD$ and $BC = BD$.

Notice that this is almost (but not quite) SSS. (Euclid will attempt to use this proposition to prove SSS, which is Proposition 8.)

Universe 3: Neutral With SAS

(a)   The proposition speaks of points C and D being on the "same side" of segment AB. Define *same side*. I'll start you off. The points C and D are on the *same side* of segment AB if...

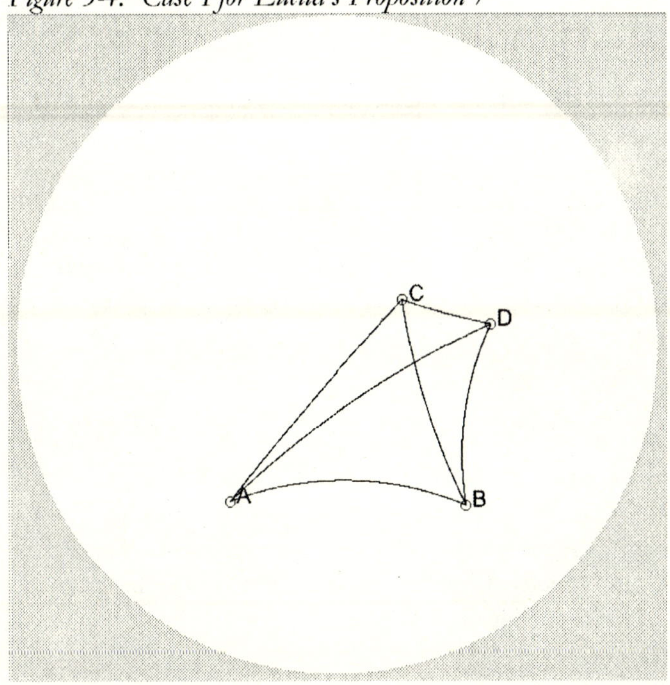

Figure 3-4: Case 1 for Euclid's Proposition 7

(b)   The picture in *The Elements* that Euclid uses to illustrate his proof is similar to Figure 3-4, except Euclid's picture is in (duh!) Euclidean Space, and not Poincaré Disk Space. However, it is actually only one of two possible cases. Draw a picture illustrating the other case. In general, when there is more than one case, Euclid only proves the most difficult one, leaving the other cases to be proved by the reader.

(c)   Fill in the blanks of the following proof of Proposition 7.

Statements	Reasons
(1) --?--	(1) RAA Hypothesis
(2) There exists a circle α with center A and radius $\overline{AC}$,	(2) --?--
(3) --?--	(3) Set Theory (or Euclid's Postulate 3).
(4) α and β intersect at points C and D	(4) --?--
(5) --?--	(5) Circular Continuity Axiom (2)
(6) Steps (1) and (5) are a contradiction.	(6) --?--

Note: Euclid's proof of this proposition in *The Elements* is fundamentally different from ours. Once again, Euclid's proof does not quite cut the mustard.

Task 17: Refer to Figure 3-5. Consider lines $\overleftrightarrow{AB}$ and $\overleftrightarrow{CD}$. Suppose the relative speed between A and C is exactly 95.6% of the speed of light. Suppose the relative speed between A and D is exactly 95.6% of the speed of light. Suppose the relative speed between B and C is exactly 95.6% of the speed of light. What can you conclude about the relative speed between B and D? Why?

Task 18: It's finally time to tackle SSS, which is Euclid's Proposition 8. Here it is in modern lingo.

> Proposition 8: Given two triangles $\triangle ABC$ and $\triangle DEF$. If $AB = DE$, $BC = EF$, and $AC = DF$, then $\triangle ABC \cong \triangle DEF$.

Fill in the blanks in Euclid's proof using the assumptions of Universe 3. If you don't have a copy of *The Elements* handy, Figure 3-6 shows a Poincaré Disk version of the picture that accompanies the proof.

Universe 3: Neutral With SAS

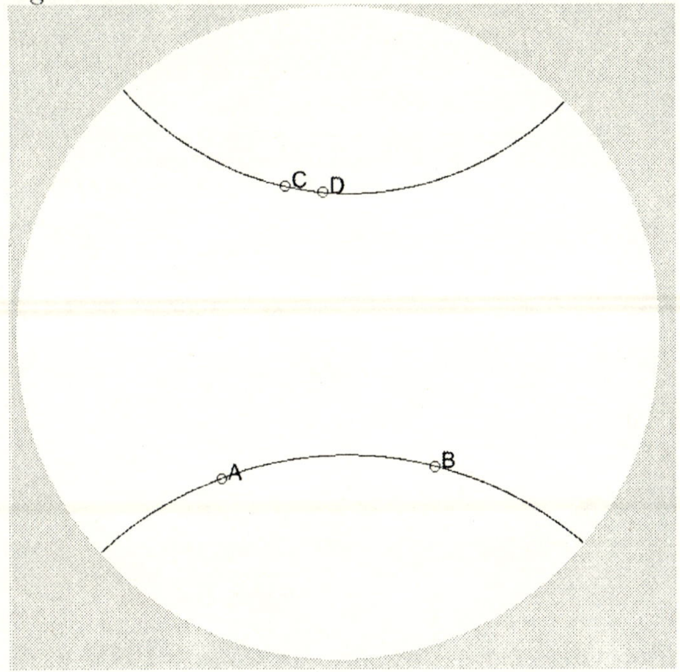

This proof is far, far better than his "proof" of SAS, and is actually pretty convincing. There is however, a fatal flaw, for there's one blank that cannot be filled in. Write "uh-oh!" in this blank.

Statements	Reasons
(a) --?-- center E and radius ED.	(a) Set Theory (or Euclid's Postulate 3).
(b) Choose G on α on the same side of $\overline{EF}$ as D, such that $\angle GEF \cong \angle ABC$.	(b) --?--
(c) --?--	(c) SAS Postulate
(d) $AB = GE$	(d) --?--
(e) --?--	(e) given
(f) $GE = DE$	(f) --?--
(g) --?--	(g) CPCTC
(h) $G = D$	(h) --?--
(i) --?--	(i) Substitution Prop, steps c and h

Figure 3-6: "Proof" of SSS.

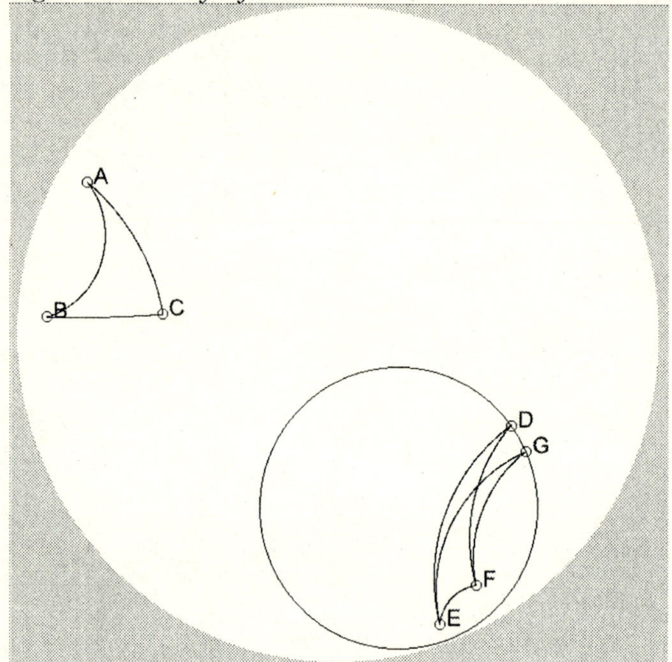

Universe 4: Neutral With SSS

In Which We Take SSS as a Postulate, and, After Proving a Few More of Euclid's Propositions, Run into Another Continuity Problem

There was a footpath leading across fields to New Southgate, and I used to go there alone to watch the sunset and contemplate suicide. I did not, however, commit suicide, because I wished to know more of mathematics.

--Bertrand Russell

Three of Euclid's Postulates

1. For every point P and for every point Q not equal to P there exists a unique line l that lies on P and Q.

2. Given segments $\overline{AB}$ and $\overline{CD}$, there exists a unique point E such that B is between A and E (or $A * B * E$), and $\overline{CD} \cong \overline{BE}$.

4. All right angles are congruent to each other.

Circular Continuity Axioms:

(1)   If a circle α with center A has one point inside and one point outside another circle β with center B, then the two circles intersect in exactly two points, and these two points are on opposite sides of $\overline{AB}$.

(2)   If circle α with center A and another circle β with center B intersect at two points, then they intersect at exactly two points, and these two points are on opposite sides of $\overline{AB}$.

Triangle Congruence Axioms

SAS Axiom: If two sides and the included angle of one triangle are congruent respectively to two sides and the included angle of another triangle, then the two triangles are congruent.

SSS Axiom: If three sides of one triangle are congruent to three sides of another triangle, then the triangles are congruent.

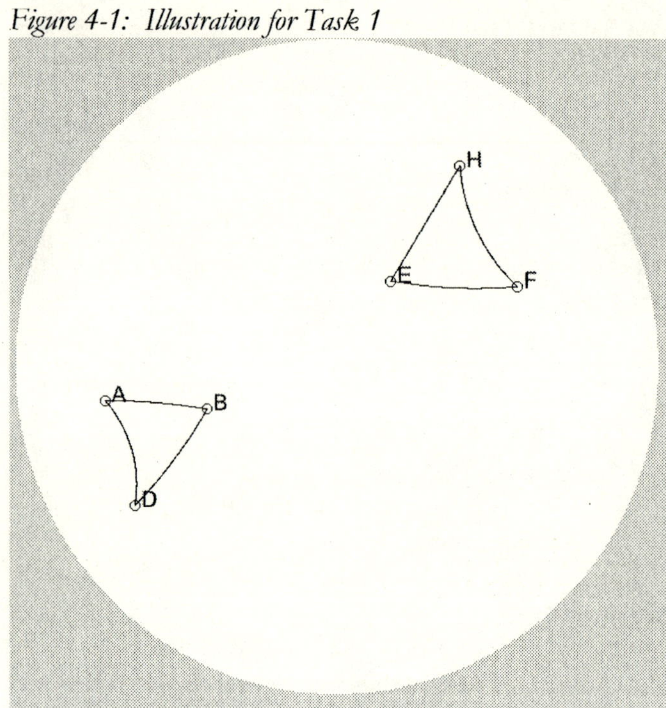

Task 1:   Refer to Figure 4-1.

A thinks that B is traveling at a velocity of (1, 0°), and that D is traveling at a velocity of (1.4, -47.6°). B thinks that A is traveling at a velocity of (1, 180°) and that D is traveling at a velocity of (1.2, -111.3°).

E thinks that F is traveling at a velocity of (1, 0°). She also thinks that H is traveling at 83.365% of the speed of light. H, meanwhile, believes that F is traveling at 88.535% of the speed of light.

Approximately what velocity does E think H is traveling? Justify your answer.

Task 2: ___   You've already proved Euclid's Proposition 1: given a segment AB, there exists a point C such that ΔABC is equilateral.

Now we will prove our first lemma. A lemma is a little theorem that needs to get proved in order to prove a major theorem (in this case, Proposition 9).

Lemma 1: Given a segment $\overline{AB}$, there exists exactly two points C and D such that $\triangle ABC$ and $\triangle ABD$ are equilateral. Moreover, C and D are on opposite sides of $\overline{AB}$.

Statements	Reasons
(1) There exists circle α with center A and radius AB, and circle β with center B and radius BA.	(1) --?--
(2) --?--	(2) Circular Continuity Axiom
(3) Since C is on α, $AB = AC$. Since C is on β, $BA = BC$.	(3) --?--
(4) --?--	(4) Definition of equilateral triangle
(5) Since D is on α, $AB = AD$. Since D is on β, $BA = BD$.	(5) --?--
(6) --?--	(6) Definition of equilateral triangle

Task 3: Draw an angle. Now bisect the angle with a straightedge and compass. If you get stuck, look at Figure 4-2.

Task 4: Define *angle bisector*. I'll start you off. "Given an angle $\angle BAC$, the ray $\overrightarrow{AF}$ is called an *angle bisector* if..."

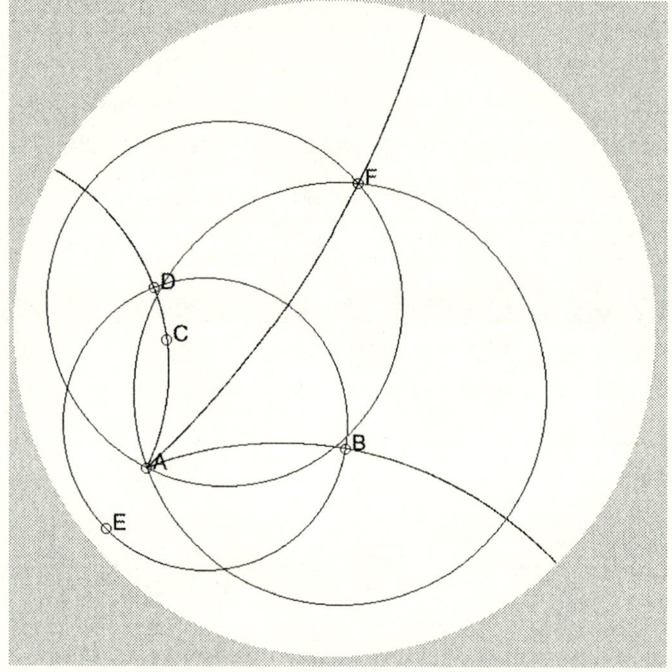

Task 5:   Here is Euclid's Proposition 9, in modern mathematical lingo.

Proposition 9: Given an angle $\angle BAC$, there exists a point F such that ray $\overrightarrow{AF}$ is an angle bisector.

And here is Euclid's proof (which can also be found—in its original form, more or less—on p. 264 of the Dover edition of *The Elements*). Here, we use modern wording, and the assumptions of Universe 4. Also, we use Lemma 1. Fill in the blanks. [Note: Euclid's proof is completely convincing, for a change!]

Statements	Reasons
(a) WMAWLOGT $AB \le AC$. Let D be a point on $\overline{AB}$. Then there exists a point E such that -- ?--.	(a) Proposition 3

(b) There exists segment $\overline{DE}$.	(b) --?--
(c) There exists a point F such that --?--.	(c) Proposition 1
(d) We may choose F so that it is on the other side of $\overline{DE}$ as A.	(d) --?--
(e) --?--	(e) SSS Postulate
(f) $\angle DAF \cong \angle EAF$	(f) --?--
(g) --?--	(g) Definition of angle bisector

Figure 4-3: $\overline{AB}$ bisected: $AE = EB$. The two circles are congruent.

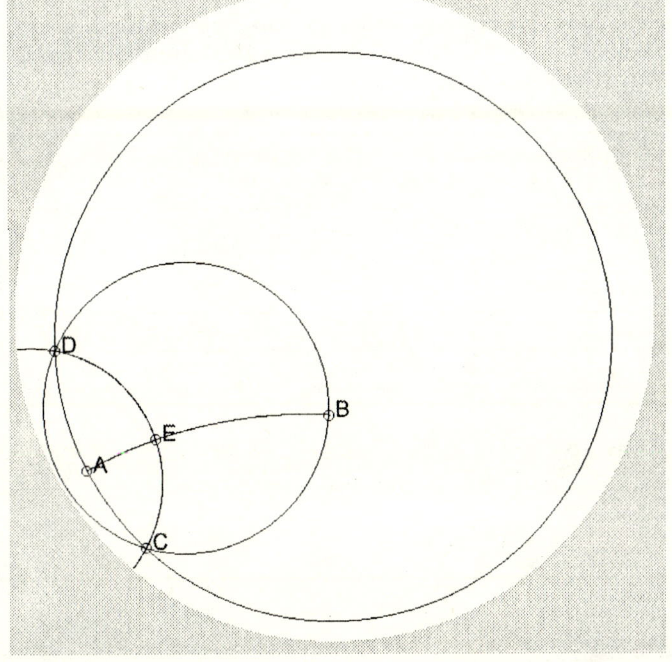

Task 6: Do you think that it is possible to trisect an angle with just a straightedge and compass?

Task 7: Draw a line segment. Now bisect it, with just a straightedge and compass. See Figure 4-3.

Task 8: Here is Euclid's Proposition 10, suitably modernized.

Proposition 10 (Midpoint Theorem): Given segment $\overline{AB}$, there exists a point D between A and B such that $AD = BD$.

Here is Euclid's proof, using modern wording, and using the assumptions of Universe 4. (Look on p. 267 of the Dover edition of *The Elements* to read a translation of the original version.) Alas, there is a fatal flaw in Euclid's proof. Put "uh-oh!" in the blank that doesn't work.

Statements	Reasons
(a) There exists a point C such that $\triangle ABC$ is equilateral.	(a) --?--
(b) There exists point E such that --?--.	(b) Proposition 9
(c) Let D be the point where ray $\overrightarrow{CE}$ intersects segment $\overline{AB}$.	(c) --?--
(d) --?--	(d) SAS Postulate
(e) $AD = BD$	(e) --?--

Task 9: In order to prove Proposition 10, there seems to be some other axiom that we need. Think of one that will serve.

Universe 5: Neutral With Even More Continuity

In Which, Despite a New Continuity Axiom, We Start to Wonder Whether Points and Lines Actually Even Exist

> *Euclid taught me that without assumptions there is no proof. Therefore, in any argument, examine the assumptions. Then, in the alleged proof, be alert for inexplicit assumptions. Euclid's notorious oversights drove this lesson home. Thanks to him, I am (I hope!) immune to all propaganda, including that of mathematics itself.*
>
> --E. T. Bell

Three of Euclid's Postulates

1. For every point P and for every point Q not equal to P there exists a unique line l that lies on P and Q.

2. Given segments $\overline{AB}$ and $\overline{CD}$, there exists a unique point E such that $A * B * E$, and $\overline{CD} \cong \overline{BE}$.

4. All right angles are congruent to each other.

Circular Continuity Axioms:

(1) If a circle α with center A has one point inside and one point outside another circle β with center B, then the two circles intersect in exactly two points, and these two points are on opposite sides of $\overline{AB}$.

(2) If circle α with center A and another circle β with center B intersect at two points, then they intersect at exactly two points, and these two points are on opposite sides of $\overline{AB}$.

Crossbar Continuity Axiom: If $\overrightarrow{AC} * \overrightarrow{AD} * \overrightarrow{AB}$, then ray $\overrightarrow{AD}$ intersects segment $\overline{BC}$. Conversely, if $C * D * B$ and A is not on line $\overleftrightarrow{CB}$, then $\overrightarrow{AC} * \overrightarrow{AD} * \overrightarrow{AB}$.

Triangle Congruence Axioms

SAS Axiom: If two sides and the included angle of one triangle are congruent respectively to two sides and the included angle of another triangle, then the two triangles are congruent.

SSS Axiom: If three sides of one triangle are congruent to three sides of another triangle, then the triangles are congruent.

Task 1:   Okay, let's try proving Proposition 10, once more, with feeling.

Proposition 10 (Midpoint Theorem): Given segment $\overline{AB}$, there exists a point D such that $A * D * B$ and $AD = BD$.

Statements	Reasons
(a) --?--	(a) Proposition 1
(b) There exists a point E such that ray $\overrightarrow{CE}$ bisects $\angle ACB$.	(b) --?--
(c) --?--	(c) Crossbar Continuity Axiom
(d) $\triangle ACD \cong \triangle BCD$	(d) --?--
(e) --?--	(f) CPCTC

Task 2:   Given a line l and a point C that lies on l, construct (with a straightedge and compass) a line through C at right angles to l. As you know, such lines are called *perpendicular*, *orthogonal*, or *normal* (depending on the textbook). But Euclid uses the term "at right angles to," so I thought I would too. See Figure 5-1 for a Poincaré "hint."

Task 3:   Here's a modern rendition of Euclid's Proposition 11.

Proposition 11: Given a line $\overleftrightarrow{AB}$ and a point C that lies on $\overleftrightarrow{AB}$, there exists a line m through C at right angles to $\overleftrightarrow{AB}$.

(We can say "m is *perpendicular* to $\overleftrightarrow{AB}$," or "$m \perp \overleftrightarrow{AB}$.")

Figure 5-1: A line through point C, perpendicular to $\overleftrightarrow{AB}$.

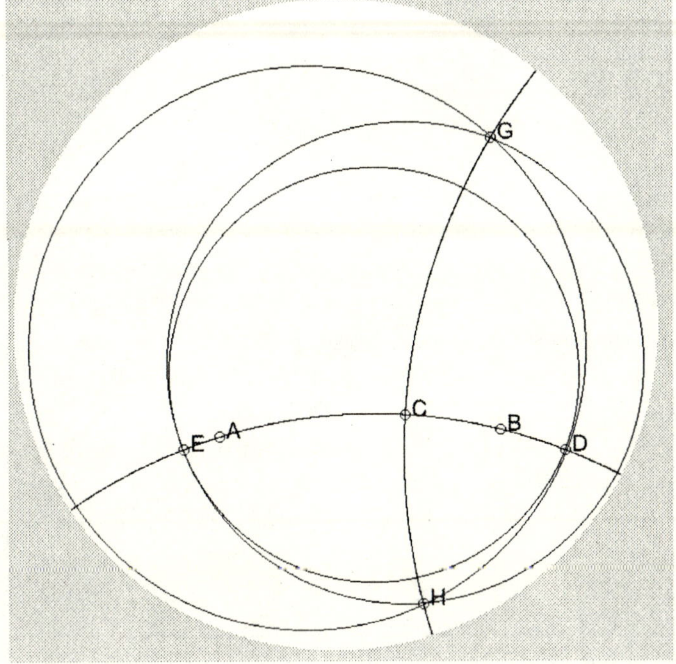

Here's Euclid's proof from *The Elements*, using modern lingo, and the assumptions of Universe 5. Fill in the blanks.

Statements	Reasons
(a) Let $D \neq C$ be a point on $\overleftrightarrow{AB}$. There exists a point E on $\overleftrightarrow{AB}$ such that $D * C * E$ and $CD = CE$.	(a) --?--
(b) --?--	(b) Proposition 1

(c) $\triangle DCG \cong \triangle ECG$	(c) --?--
(d) --?--	(d) CPCTC
(e) $\angle DCG$ and $\angle ECG$ are supplementary angles	(e) --?--
(f) --?--	(f) Definition of "right angles"

Task 4: In Universe 5, can you think of any way to prove that for every line *l*...

    (a)   there exists a point lying on *l* ?

    (b)   there exists a point not lying on *l* ?

Task 5: In Universe 5, can you think of any way to prove that the plane is nonempty, that is, that points and lines exist?

Task 6: Statements that can be proved are called *theorems*. So far we've plowed through the first eleven of Euclid's propositions. All of them are true statements in Universe 5. Additionally, all of them are *theorems* in Universe 5, except Proposition 4 (SAS) and Proposition 8 (SSS). SAS and SSS are both un-provable and un-debunkable given our other postulates and axioms, and yet we want them to be true; hence we *declare* them to be true. We take them as *axioms* (which are also known as *postulates*). However, we didn't *have* to take them as axioms. We could have simply created a universe where SAS and SSS are *not* necessarily true. Indeed, in Universe 1, SAS and SSS are not necessarily true. Since mathematics is the study of all possible universes, exploring Universe 1 is completely legal and safe. Have fun!

However, suppose you find a statement in a certain universe that nobody has found a proof for, but nobody has found a counterexample for either. In other words, nobody yet knows whether or not the statement is true or false in this universe. You would rather not take this statement as an axiom, and yet you *do* want it to be true. (Maybe it has some interesting consequences that would make the universe more fun.) Here's the question: is it possible that the statement *is* true, but no proof (or counterexample) for it even exists? In other words, in a given universe, do we know that all true

Universe 5: Neutral With Even More Continuity

statements can be proved? Or is it possible to have a statement that is true, and yet no proof for it is possible?

Task 7:   Some questions to ponder.

(a)   Do you think that the axiomatic method can be applied to subjects other than mathematics?

(b)   Can the U.S. Constitution (including all its amendments) be considered the list of postulates and axioms from which the federal courts deduce all rules of law?

(c)   Do you think the "truths" asserted in the Declaration of Independence are "self-evident"?   For reference, consider the following list of "axioms":

> (i)   All men are created equal.

> (ii)   They are endowed by their Creator with certain unalienable Rights, and that among these are Life, Liberty and the pursuit of Happiness.

> (iii)   To secure these rights, Governments are instituted among Men, deriving their just powers from the consent of the governed.

> (iv)   Whenever any Form of Government becomes destructive of these ends, it is the Right of the People to alter or to abolish it, and to institute new Government, laying its foundation on such principles and organizing its powers in such form, as to them shall seem most likely to effect their Safety and Happiness.

Task 8:   Define:

(a)   The *midpoint* of segment $\overline{AB}$. ("The *midpoint* of segment $\overline{AB}$ is a point M such that..."

(b)   The sides *opposite to* and *adjacent to* a given vertex A of triangle ΔABC. ("Given ΔABC, the side *opposite to* vertex A is...")

(c)   The *medians* of a triangle. ("A *median* of a triangle is...")

(d)   The *altitudes* of a triangle. ("An *altitude* of ΔABC is...")

(e)   A *right triangle.*

Task 9:   Define the *quadrilateral ABCD*. Note that the order in which the letters are written is essential. For instance, just because $ABCD$ is a quadrilateral, it doesn't mean that $ACBD$ is one. Also, if $ACBD$ is a quadrilateral, it would not be the same one as $ABCD$.

While you're at it also define the *sides* and the *vertices* of a quadrilateral. I'll start you off. "Given four points A, B, C, and D, no three of which are collinear, and such that..."

Task 10:   Using your definitions in the previous task, define:

(a)   *adjacent* sides of quadrilateral $ABCD$.

(b)   *opposite* sides of quadrilateral $ABCD$.

(c)   a *diagonal* of quadrilateral $ABCD$.

(d)   a *parallelogram.* (Use the word *parallel.*)

Universe 6: Neutral with Incidence
In Which We Finally Prove that Points and Lines Actually Exist

> *Let no one ignorant of geometry enter this door.*
> --Entrance to Plato's Academy

Hilbert's Axioms of Incidence

IA-1   For every point P and for every point Q not equal to P there exists a unique line l incident with P and Q.

IA-2   For every line l there exists at least two distinct points that are incident with l.

IA-3   There exists three distinct points with the property that no line is incident with all three of them.

Two of Euclid's Postulates

E-2   Given segments $\overline{AB}$ and $\overline{CD}$, there exists a unique point E such that $A * B * E$, and $\overline{CD} \cong \overline{BE}$.

E-4   All right angles are congruent to each other.

Circular Continuity Axioms:

(1)   If a circle α with center A has one point inside and one point outside another circle β with center B, then the two circles intersect in exactly two points, and these two points are on opposite sides of $\overline{AB}$.

(2)   If circle α with center A and another circle β with center B intersect at two points, then they intersect at exactly two points, and these two points are on opposite sides of $\overline{AB}$.

Crossbar Continuity Axiom: If $\overrightarrow{AC} * \overrightarrow{AD} * \overrightarrow{AB}$, then ray $\overrightarrow{AD}$ intersects segment $\overline{BC}$. Conversely, if $C * D * B$ and A is not on line $\overleftrightarrow{CB}$, then $\overrightarrow{AC} * \overrightarrow{AD} * \overrightarrow{AB}$.

Triangle Congruence Axioms

SAS Axiom: If two sides and the included angle of one triangle are congruent respectively to two sides and the included angle of another triangle, then the two triangles are congruent.

SSS Axiom: If three sides of one triangle are congruent to three sides of another triangle, then the triangles are congruent.

Task 1: In this universe, Hilbert's three axioms of incidence replace Euclid's first postulate. Reading through them, what do you think the word *incident* means? (As I mentioned before, during the first quarter of the twentieth century, the leading mathematician in the world was David Hilbert. His name pops up in many, many fields of mathematics—and even in physics—but here we are interested in his contributions to the foundations of geometry. He completely axiomized Euclidean geometry in a rigorous way. At around the same time, Bertrand Russell was working on the foundations of arithmetic, so it was a big era of mathematical spring-cleaning. Russell, along with another mathematician named Whitehead, came out with a tome called *Principia Mathematica* in which, starting with some basic axioms of logic, he "created" the counting numbers. After about 240 pages they finally prove (in excruciating detail) that $1 + 1 = 2$. It makes for incredibly boring reading, and yet, I'm glad that somebody did it.)

Task 2: In Universe 6, prove that for every line l

(a) there exists a point lying on l.

(b) there exists a point not lying on l.

Task 3: In Universe 6, prove that the plane is nonempty, i.e., prove that points and lines exist.

Task 4: Complete the definition: "Lines *l* and *m* are called *parallel* if..."

Task 5: We will now prove that if *l* and *m* are distinct lines that are not parallel, then *l* and *m* have a unique point in common. Fill in the blanks in the two-column proof below.

Statement	Reason
(1) *l* and *m* are distinct lines	(1) --?--
(2) --?-- have a point P in common	(2) definition of "not parallel"
(3) Suppose, to the contrary, that there exists another point $Q \neq P$ that lies on *l* and *m*	(3) --?--
(4) --?--	(4) IA-1
(5) Steps 1 and 4 contradict each other	(5) --?--

Task 6: Three distinct lines are called *concurrent* if they have a unique point in common. We will now prove that there exists three distinct lines that are not concurrent. Fill in the blanks in the two-column proof below.

Statement	Reason
(1) There exists three points, $P, Q,$ and R, such that no line is incident with all three of them.	(1) --?--
(2) --?-- $\overleftrightarrow{PQ}, \overleftrightarrow{QR},$ and $\overleftrightarrow{PR}$.	(2) IA-1
(3) Suppose, to the contrary, that these three lines are concurrent, and meet at a point S.	(3) --?--
(4) Since $\overleftrightarrow{PQ}$ and $\overleftrightarrow{QR}$ have point Q in common, --?--	(4) Task 5, above.

Statement	Reason
(5) Since $\overleftrightarrow{PQ}$ and $\overleftrightarrow{PR}$ have point P in common, $P = S$.	(5) --?--
(6) --?--	(6) Steps 4 and 5, and the Substitution Property.
(7) Steps 6 and 1 contradict each other.	(7) --?--

Task 7:   We will now prove that for every point P there exists at least two lines through P. Fill in the blanks in the two-column proof below.

Statement	Reason
(1) There exists three points Q, R, and S such that no line is incident with all three of them.	(1) --?--
(2) If P is distinct from the points Q, R, and S, then --?--. If P is the same as one of the points, say Q, then --?--	(2) IA-1
(3) Lines $\overleftrightarrow{PQ}$ (if it exists), $\overleftrightarrow{PR}$, and $\overleftrightarrow{PS}$ can't all be the same line.	(3) --?--

Task 8:   We will now prove that for every point P there exists at least one line not passing through it.

Statement	Reason
(1) --?-- Q, R, and S --?--	(1) IA-3
(2) Lines $\overleftrightarrow{PQ}$, $\overleftrightarrow{PR}$, and $\overleftrightarrow{PS}$ can't all be the same line, so we may assume $\overleftrightarrow{PS}$ is a different line.	(2) --?--
(3) --?-- $\overleftrightarrow{SR}$.	(3) IA-1
(4) We claim P does not lie on $\overleftrightarrow{SR}$. Suppose, to the contrary, that it does.	(4) --?--

Universe 6: Neutral With Incidence

	(5) Task 5, Universe 6.
(5) Then $\overleftrightarrow{PS}$ and $\overleftrightarrow{SR}$ --?--	
(6) Similarly, $\overleftrightarrow{PR}$ and $\overleftrightarrow{SR}$ would have two points in common, which would mean that $\overleftrightarrow{PR} = \overleftrightarrow{SR}$.	(6) --?--
(7) --?--	(7) Steps 5 and 6, and substitution.
(8) Steps 7 and 2 are a contradiction.	(8) --?--

Task 9: Consider the set $\{A, B, C\}$ of three letters, which we will call "points." "Lines" will be those subsets that contain exactly two letters: $\{A, B\}$, $\{A, C\}$, and $\{B, C\}$. A "point" will be interpreted as "incident" with a "line" if it is a member of that subset. Thus, under this interpretation, A lies on $\{A, B\}$ and $\{A, C\}$, but does not lie on $\{B, C\}$.

(a)   In this interpretation, do parallel lines exist?

(b)   Is this interpretation a model for Incidence Geometry? (In other words, are all the Axioms of Incidence satisfied?) If not, which axiom(s) fail(s)?

(c)   Suppose that you answered "yes" to part (b). Would that mean that all the results that you proved in Tasks 5 through 8 would have to also be true for this interpretation?

Task 10:   Suppose we interpret "points" as points on a sphere, "lines" as great circles on the sphere, and "incidence" in the usual sense, as a point lying on a great circle.

(a)   In this interpretation, do parallel lines exist?

(b)   Is this interpretation a model for Incidence Geometry? If not, which axiom(s) fail(s)?

Task 11: Given a line l, and a point C which is not on l. With a straightedge and compass, construct a line through C perpendicular to l. [Hint, sort of: see Figure 6-1.]

Figure 6-1: A line through C perpendicular to line $\overleftrightarrow{AB}$.

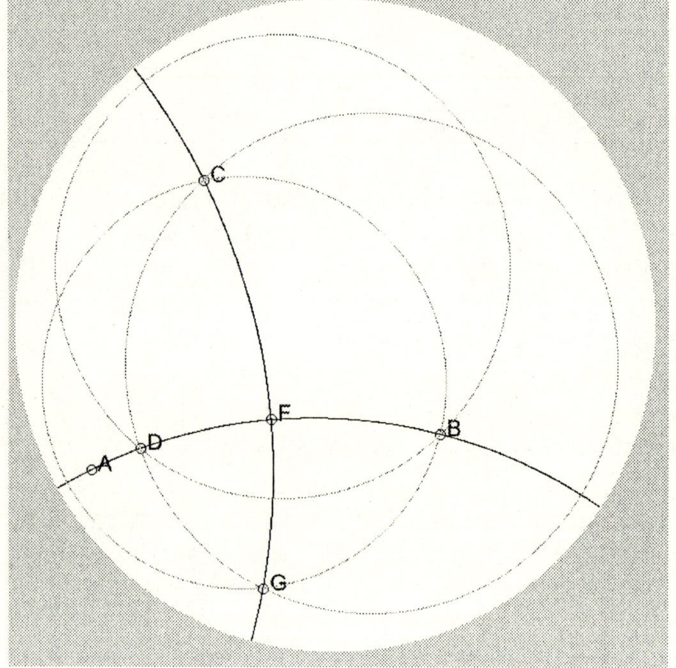

Task 12: Here is a modern rendition of Euclid's Proposition 12.

Proposition 12: Given a line $\overleftrightarrow{AB}$ and a point C that does not lie on $\overleftrightarrow{AB}$. Then there exists a line through C perpendicular to $\overleftrightarrow{AB}$.

Fill in the blanks in the two-column proof. Yes, yes, I know that we've only just begun a brand new universe, and that we're all ready to settle down into our nice comfy tour of Euclid's *Elements*, but—alas!—there is a problem with this proof. Oh well. That's life in the big city. Put "uh-oh!" in the offending space.

Statements	Reasons
(a) If $\overleftrightarrow{CB} \perp \overleftrightarrow{AB}$, then we're done. So WMAWLOGT they're not perpendicular. There exists --?--	(a) Set Theory (or Euclid's Postulate 3)
(b) $\overleftrightarrow{AB}$ intersects circle α at exactly two points, B and D	(b) --?--
(c) --?--	(c) Proposition 10
(d) $\triangle DFC \cong \triangle BFC$	(d) --?--
(e) --?--	(e) CPCTC, or definition of congruent triangles
(f) $\angle DFC$ and $\angle BFC$ are supplementary	(f) --?--
(g) --?--	(g) steps e and f, and definition of "right angle"

81

Universes 7 a, b, c, and d: Incidence Geometry
In Which We Discover What Puts the "Euclid" into Euclidean Geometry

> *Out of nothing, I have created a strange new universe.*
> --János Bolyai, co-discoverer of Hyperbolic Geometry

Parallel Postulates (Choose Only One, Please)

(a)   Euclidean: For every line l and for every point P that does not lie on l, there exists a unique line m through P that is parallel to l.

(b)   Hyperbolic: There exists a line l and a point P not lying on l such that at least two distinct lines parallel to l pass through P.

(c)   Elliptic: Parallel lines do not exist.

(d)   None of the above!

Hilbert's Axioms of Incidence

IA-1   For every point P and for every point Q not equal to P there exists a unique line l incident with P and Q.

IA-2   For every line l there exists at least two distinct points that are incident with l.

IA-3   There exists three distinct points with the property that no line is incident with all three of them.

Task 1:   Hello! Welcome to *Whose "Line" is It Anyway?* This is the show where the postulates are made up, and the "points" don't matter. That's right, the "points" are just like tables, chairs, and beer mugs. (Cue laugh track.) In our first game, let "points" be the four letters in the set $\{A, B, C, D\}$. Let "lines" be all six sets containing exactly two of these letters. Let "incidence" be set membership, as in Task 9 of Universe 6.

                                        Universe 7: Incidence Geometry

(a)   Is this interpretation a model for Incidence Geometry?   If not, which axiom(s) fail(s)?

(b)   Does this interpretation have the Euclidean, hyperbolic, or elliptic parallel property?

(c)   How does Figure 7-1 relate to this model?

Figure 7-1: Illustration for Task 1

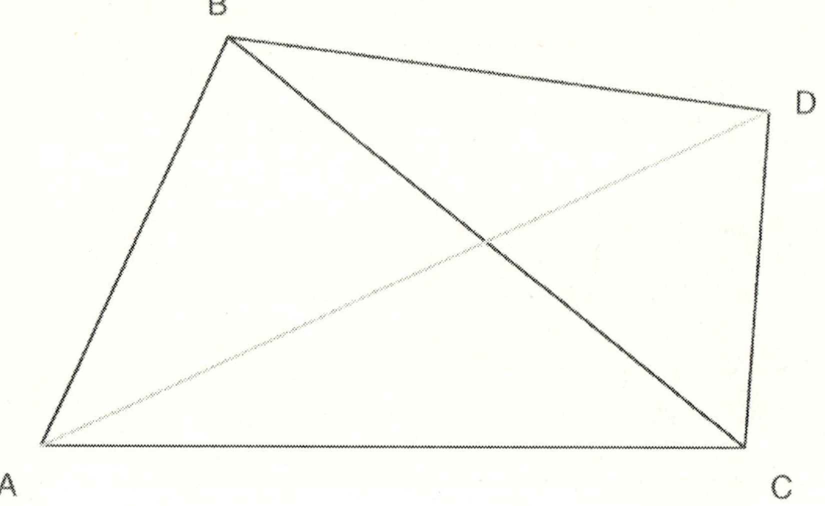

Task 2:   Let "points" be the five letters in the set {*A, B, C, D, E*}. Let "lines" be all ten sets containing exactly two of these letters. Let "incidence" be set membership.

(a)   Is this interpretation a model for Incidence Geometry?   If not, which axiom(s) fail(s)?

(b)   Does this interpretation have the Euclidean, hyperbolic, or elliptic parallel property?

(c)   How does Figure 7-2 relate to this model?

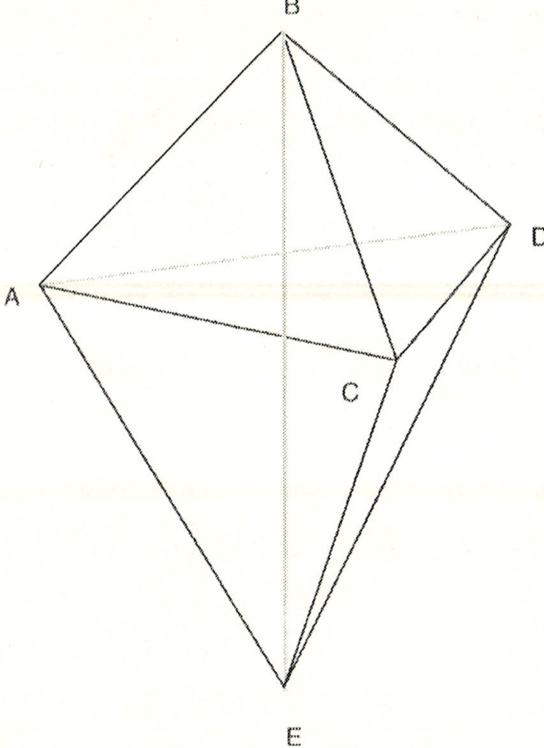

Task 3: Suppose "points" are dots on a sheet of paper, "lines" are circles drawn on the paper, and "incidence" means that the dot lies on the circle.

(a) Is this interpretation a model for Incidence Geometry? If not, which axiom(s) fail(s)?

(b) Does this interpretation have the Euclidean, hyperbolic, or elliptic parallel property?

Task 4: Suppose "points" are lines in three-dimensional space, "lines" are planes in three-dimensional space, and "incidence" is the usual relation of a line lying in a plane.

(a) Is this interpretation a model for Incidence Geometry? If not, which axiom(s) fail(s)?

(b)   Does this interpretation have the Euclidean, hyperbolic, or elliptic parallel property?

Task 5:   Suppose we're given a fixed point P in three-dimensional space. "Points" are lines that go through P, "lines" are planes that go through P, and "incidence" is the usual relation of a line lying in a plane.

(a)   Is this interpretation a model for Incidence Geometry? If not, which axiom(s) fail(s)?

(b)   Does this interpretation have the Euclidean, hyperbolic, or elliptic parallel property?

Task 6:   Fix a circle on the plane. Interpret "point" to mean an ordinary point inside the circle, interpret "line" to mean a chord of the circle, and let "incidence" mean that the point lies on the chord in the usual sense. (A chord of a circle is a segment whose endpoints lie on the circle.)   This model is called the *Klein Disk*, after the mathematician Felix Klein (1849-1925). This is the same Klein that is responsible for the "Klien bottle," a four-dimensional version of the Möbius strip.

(a)   Is this interpretation a model for Incidence Geometry? If not, which axiom(s) fail(s)?

(b)   Does this interpretation have the Euclidean, hyperbolic, or elliptic parallel property?

Task 7:   Fix a sphere in three-dimensional space. Two points on a sphere are called *antipodal* if they lie on the diameter of the sphere; e.g., the north and south poles are antipodal. Interpret "point" to be a set $\{P_1, P_2\}$ consisting of two antipodal points on the sphere. Interpret "line" to be a great circle α on the sphere. Interpret a "point" $\{P_1, P_2\}$ to "lie on" a "line" α if one of the points P_1 or P_2 lies on the great circle α (in which case the other point also lies on α). This model is called the *Riemann Sphere*, after the mathematician Bernhard Riemann (1826-1866), who eventually developed a non-Euclidean Geometry that was adopted by Albert

86

Einstein in his General Theory of Relativity. (This is the same Riemann who is responsible for the *Riemann Integral* in calculus. He was a busy guy.)

(a)   Is this interpretation a model for Incidence Geometry? If not, which axiom(s) fail(s)?

(b)   Does this interpretation have the Euclidean, hyperbolic, or elliptic parallel property?

Task 8:   Construct a model of Incidence Geometry that has neither the elliptic, hyperbolic, nor Euclidean parallel properties. [Hint: These properties refer to any line l and any point P not on l. Construct a model that has different parallelism properties for different choices of l and P. Five points suffice.]

Task 9:   The "Euclidean parallel postulate" given at the beginning of this universe is sometimes called *Playfair's Postulate* (because it appeared in John Playfair's geometry textbook published in 1795):

> For every line l and for every point P that does not lie on l, there exists a unique line m through P that is parallel to l.

Now this is different than the original postulate given by Euclid in *The Elements*. Here is a modern rewording of Euclid's original parallel postulate:

> If a line t intersects lines m and n such that a pair of same-side interior angles are less than the sum of two right angles, then m and n are *not* parallel, and in fact will intersect on the same side of t as these same-side interior angles.

In Universe 9, we will show that these two postulates—Playfair's and Euclid's original—are logically equivalent. Why do you suppose that modern mathematicians favor Playfair's version over Euclid's original version?

     Universe 7: Incidence Geometry

Task 10: Suppose that in a given model for Incidence Geometry, every "line" has at least three distinct "points" lying on it.

(a) What are the least number of "points" and the least number of "lines" that such a model can have?

(b) Suppose further that the model has the Euclidean parallel property. Show that 9 is now the least number of "points" and 12 the least number of "lines" such a model can have.

Task 11: For each pair of the Axioms of Incidence, construct an interpretation in which those two axioms are satisfied but the third is not. (This will show that the three axioms are *independent*, in the sense that it is impossible to prove any one of them from the other two.)

Task 12: Consider Einstein's Special Theory of Relativity, where "points" are rapidity velocities in polar coordinates, and "lines" are the "lines" of a Poincaré Disk.

(a) Is this interpretation a model for Incidence Geometry? If not, which axiom(s) fail(s)?

(b) Does this interpretation have the Euclidean, hyperbolic, or elliptic parallel property?

Universe 8: Complete Neutral Geometry

*In Which We Finally Fix All of Euclid's Continuity and Existence Problems,
and Dwell Contentedly in the Present Moment*

*I believe that mathematical reality lies outside us, that our function is discover or observe
it, and that the theorems which we prove, and which we describe grandiloquently as our
"creations," are simply notes of our observations.*

--G. H. Hardy

Hilbert's Axioms of Incidence

IA-1   For every point P and for every point Q not equal to P there exists a unique line l incident with P and Q.

IA-2   For every line l there exists at least two distinct points that are incident with l.

IA-3   There exists three distinct points with the property that no line is   incident with all three of them.

Two of Euclid's Postulates

E-2   Given segments $\overline{AB}$ and $\overline{CD}$, there exists a unique point E such that $A * B * E$, and $\overline{CD} \cong \overline{BE}$.

E-4   All right angles are congruent to each other.

Crossbar Continuity Axiom: If $\overrightarrow{AC} * \overrightarrow{AD} * \overrightarrow{AB}$, then ray $\overrightarrow{AD}$ intersects segment $\overline{BC}$.   Conversely, if $C * D * B$ and A is not on line $\overleftrightarrow{CB}$, then $\overrightarrow{AC} * \overrightarrow{AD} * \overrightarrow{AB}$.

Circular Continuity Axioms:
(1)   If a circle α with center A has one point inside and one point outside another circle β with center B, then the two circles intersect in exactly two points, and these two points are on opposite sides of $\overline{AB}$.

(2)   If circle α with center A and another circle β with center B intersect at two points, then they intersect at exactly two points, and these two points are on opposite sides of $\overline{AB}$.

Elementary Continuity Axiom:

Suppose a circle with center O intersects a line at point A. If $\overleftrightarrow{OA}$ is perpendicular to the line, then the circle is tangent to the line. Otherwise, the circle intersects the line at exactly two points.

Triangle Congruence Axioms

SAS Axiom: If two sides and the included angle of one triangle are congruent respectively to two sides and the included angle of another triangle, then the two triangles are congruent.

SSS Axiom: If three sides of one triangle are congruent to three sides of another triangle, then the triangles are congruent.

Task 1:   Hello! Welcome to Complete Neutral Geometry, which is equivalent to both Euclidean and Hyperbolic Geometry, except that we omit those pesky little parallel postulates. Euclid's first 28 propositions are all theorems in Neutral Geometry, so it should be smooth sailing for a while (before all heck breaks loose with Proposition 29, that is).

So let's try proving Euclid's Proposition 12, once more, with feeling.

Proposition 12: Given a line $\overleftrightarrow{AB}$ and a point C that does not lie on $\overleftrightarrow{AB}$. Then there exists a line through C perpendicular to $\overleftrightarrow{AB}$.

Fill in the blanks in the two-column proof.

                                   Universe 8: Complete Neutral Geometry

Statements	Reasons
(a) If $\overrightarrow{CB} \perp \overrightarrow{AB}$, then we're done. So WMAWLOGT they're not perpendicular. There exists -- ?--.	(a) Set Theory (or Euclid's Postulate 3)
(b) $\overrightarrow{AB}$ intersects circle α at exactly two points, B and D	(b) --?--
(c) --?--	(c) Proposition 10
(d) $\Delta DEC \cong \Delta BEC$	(d) --?--
(e) --?--	(e) CPCTC, or definition of congruent triangles
(f) $\angle DEC$ and $\angle BEC$ are supplementary	(f) --?--
(g) --?--	(g) steps e and f, and definition of "right angle"
(h) $\overleftrightarrow{CE} \perp \overleftrightarrow{AB}$	(i) --?--

Task 2: Did you need to invoke the Euclidean parallel postulate for the proof of Proposition 12?

Task 3: Is Proposition 12 true in a hyperbolic universe (that is, one in which the hyperbolic parallel postulate is true)?

Task 4: Recall the Riemann Sphere. Use this model to explain why, in an elliptic universe, the perpendicular described by Euclid's Proposition 12 is not necessarily unique. (In a Euclidean or hyperbolic universe, the perpendicular *is* unique.)

Task 5: Recall that there are five undefined terms: *point, line, congruent, between,* and *lies on.* One of these undefined terms doesn't make sense in elliptic geometry. Which one?

Task 6: Which postulates or axioms of Universe 8 don't work for elliptic geometry?

91

Task 7: Since all right angles are congruent by E-4, they all have the same angle measure, such as 90°, or π radians. It is intuitively obvious that therefore the sum of all pairs of supplementary angles should be 180°, or 2π. Nevertheless, Euclid is quite right in feeling that this fact needs to be proved. (We have already seen that there are things that are counterintuitive that are nonetheless true.) Here, then, is a modern rendition of Euclid's Proposition 13.

Proposition 13: Given a line $\overleftrightarrow{AB}$ and another line $\overleftrightarrow{CD}$ that intersects $\overleftrightarrow{AB}$ at the point B. We assume that $C*B*D$.

Consider the angles $\angle ABD$ and $\angle ABC$. Then either both of these angles are right angles, or they both add up to two right angles. (In other words, $m\angle ABD + m\angle ABC = \pi$.)

Fill in the blanks in the following proof.

Statements	Reasons
(a) If $\angle ABD \cong \angle ABC$, then they are both right angles, and we're done. Hence, we may assume that the two angles are not congruent.	(a) --?--
(b) --?--	(b) Proposition 11
(c) We may assume that E lies on the same side of $\overleftrightarrow{CD}$ as A.	(c) --?--
(d) --?--	(d) Definition of perpendicular
(e) $m\angle CBE + m\angle DBE = m\angle CBD$	(e) --?--
(f) --?--	(f) Definition of Angle Addition
(g) $m\angle CBE + m\angle DBE = m\angle CBA + m\angle DBA$	(g) --?--

Task 8: Suppose that $\angle ABE \cong \angle ABD$. Does it necessarily follow that $\angle ABE = \angle ABD$?

Task 9:   Here is a modern rendition of Euclid's Proposition 14. We're going to depart from Euclid's practice of measuring all angles in terms of right angles. Instead, we're going to use radian measure, because it's easier, and life is short.

Proposition 14:   Let the line $\overleftrightarrow{AB}$ be given. Suppose the points C and D lie on opposite sides of $\overleftrightarrow{AB}$. If the $m\angle ABC + m\angle ABD = \pi$, then the points C, B, and D are collinear.

(a)   Sketch a picture that illustrates this proposition.

(b)   The following proof comes from Euclid's Elements. Fill in the blanks.

Statments	Reasons
(a) There exists a point E such that $C * B * E$, and $BE = BD$.	(a) --?--
(b) --?--	(b) Proposition 13
(c) $m\angle ABC + m\angle ABD = \pi$	(c) --?--
(d) --?--	(d) Substitution Property, steps b and c
(e) $m\angle ABE = m\angle ABD$	(e) --?--
(f) --?--	(f) SAS Postulate
(g) $AE = AD$	(g) --?--
(h) --?--	(h) Proposition 7
(i) C, B, and D are collinear	(i) --?--

Task 10:   See Figure 8-1.

(a)   Define *vertical angles*. I'll start you off. "A pair of angles $\angle AEC$ and $\angle BED$ are called *vertical angles* if..."

(b)   Name another pair of vertical angles in Figure 8-1.

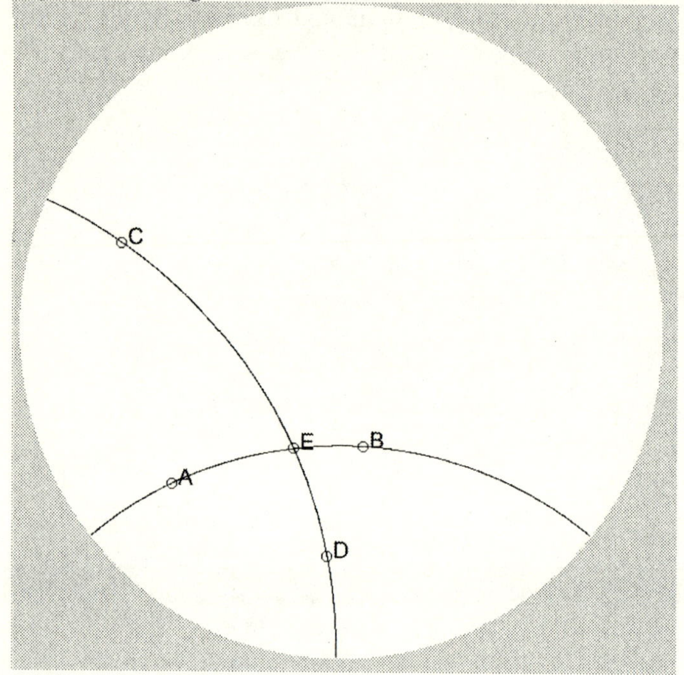

Task 11:   Euclid's Proposition 15 is the famous Vertical Angle Theorem.

Proposition 15: Vertical angles are congruent.

Fill in the blanks in the following proof.

Statements	Reasons
(a) Suppose $\angle CEA$ and $\angle DEB$ are a pair of vertical angles, with $\overrightarrow{EC}$ and $\overrightarrow{ED}$ as a pair of opposite rays. Then $m\angle AED + m\angle CEA = \pi$.	(a) --?--
(b) --?--	(b) Proposition 13
(c) $m\angle AED + m\angle CEA$ $= m\angle AED + m\angle DEB$	(c) --?--
(d) --?--	(d) Subtraction Property

Task 12:   Refer to Figure 8-2. *C* thinks that *D* is traveling at a velocity of (2.926, 0°), that *E* is traveling at a velocity of (2.319, 317.9°), and that *A* is traveling at 86.172% of the speed of light. *B* thinks that *C* is traveling at a velocity of (1.515, 0°), that *A* is traveling at a rapidity of 1.3, and that *F* is going at 89.167% of the speed of light. What velocity does *B* think that *F* is traveling?

Figure 8-2: Illustration for Task 12

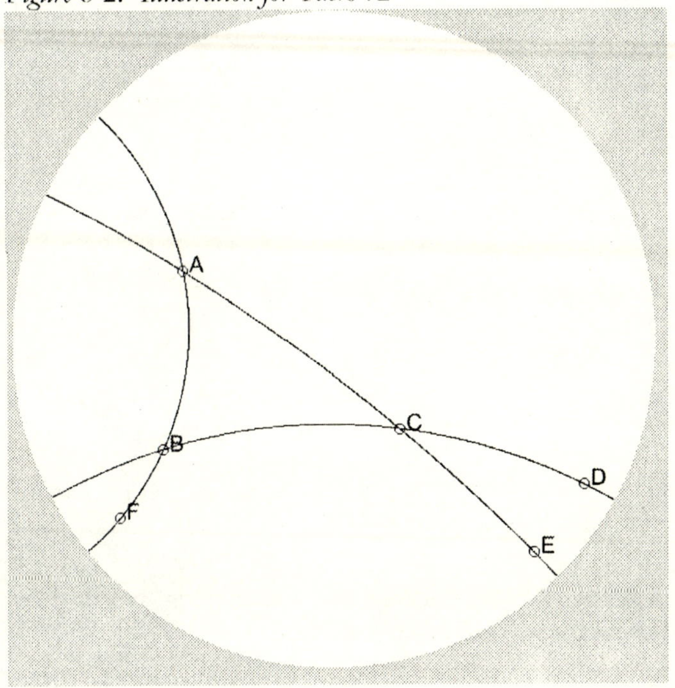

Task 13:   Define *exterior angle* of a triangle. See Figure 8-3. I'll start you off. "Given a triangle Δ*ABC*, ..." Also define the *remote interior angles*.

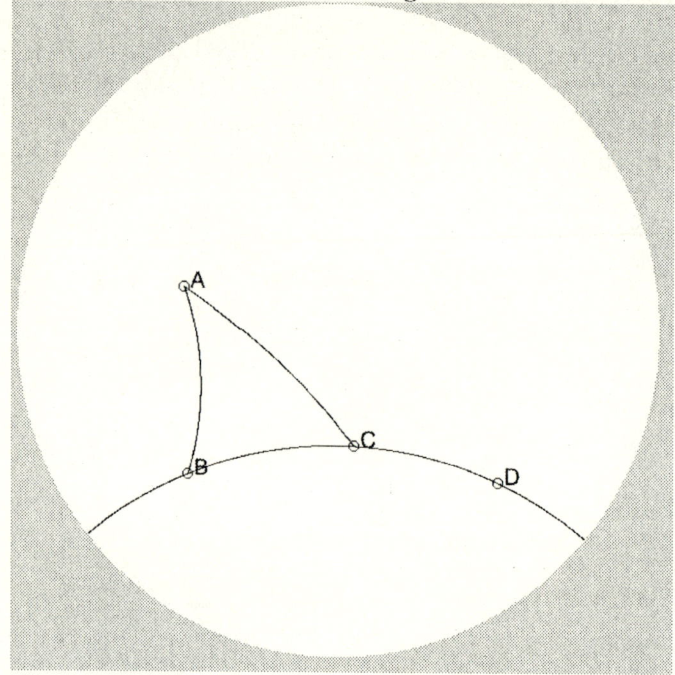

Figure 8-3: $\angle ACD$ is an exterior angle to $\triangle ABC$.

Task 14: Euclid's Proposition 16 is another familiar theorem from high school geometry.

> Proposition 16: An exterior angle of a triangle is greater than either of the remote interior angles.

Fill in the blanks of the following proof, due to Euclid. Note: I "zoomed out" a bit, and skipped a few obvious steps. I also truncated the proof, by proving it for only one of the remote angles. (Euclid does the same thing.) The proof for the other remote angle is similar.

Statements	Reasons
(a) Given triangle $\triangle ABC$ with exterior angle $\angle ACD$. We claim that $\angle ACD$ is greater than $\angle BAC$. Let segment $\overline{AC}$ be bisected at the point E.	(a) --?--

Universe 8: Complete Neutral Geometry

(b) Let F be on the line $\overleftarrow{BE}$ such that --?--.	(b) Postulate E-2
(c) $\angle AEB \cong \angle FEC$	(c) --?--
(d) --?--	(d) SAS Postulate
(e) $\angle BAE \cong \angle FCE$	(e) --?--
(f) --?--	(f) Definition of angle measure
(g) $m\angle ACD > m\angle BAC$	(g) --?--

Task 15: Euclid's Proposition 16 is true in Euclidean and Hyperbolic universes, but not in Elliptic universes. The place where Euclid's proof falls apart in elliptic space is with the assumption that ray $\overrightarrow{CF}$ must be between rays $\overrightarrow{CA}$ and $\overrightarrow{CD}$. Use the Riemann model of elliptic geometry to explain how this assumption could fall apart. Hint: First create a triangle that violates Proposition 16. (If you can find a globe, it will help you to visualize what's going on.)

Task 16: Suppose that you're a two-dimensional creature who lives in a flat surface. Suppose you're like a flat mermaid or merman, and your universe is like a flat ocean that you can swim around in. Suppose that you have a smart brain, and you suspect that your universe is actually not a flat plane, but the surface of a large sphere. Of course, you don't use the word "sphere" when discussing this possibility, for such things are quite unknown in your world. You use the term "hyper-circle;" but it means the same thing. In other words, your universe is "curved" in a third dimension. Since you're a flat two-dimensional creature, and you've always been a flat two-dimensional creature, you have a hard time imagining exactly what a "third-dimension" is, and you have even a harder time explaining it to your fellow mer-people. How would you try to convince them that their space really is curved in a mythological third-dimension?

Task 17: In actuality, you are really, of course a three-dimensional creature. (Yes, yes, I know that time can be considered a fourth-dimension; don't try to teach your grandma to suck eggs, buster. I'm talking about *spatial* dimensions here.) What might lead you to suspect that your universe is not "flat", but actually "curved" in a fourth dimension?

Universe 8: Complete Neutral Geometry

Task 18: We're back in Flatland. Suppose your universe is actually the surface of a cylinder. Could you still figure out that your space is curved in a third dimension by drawing a triangle? Hint: If you draw a triangle on a flat sheet of paper, and then roll the paper into a cylinder, does this change the angle measures in the triangle?

Task 19: We're back in our own three-dimensional universe. Suppose that you successfully manage to look at lots of huge triangles, and none of them violate Proposition 16. Does that mean that our space isn't curved in a fourth-dimension?

Task 20: Here is a modern rendition of Euclid's Proposition 17.

> Proposition 17: In any triangle, the sum of the measures of any   two angles is less than π.

Here is a modern version of Euclid's proof. Actually, Euclid only proves one of three cases. However, the proofs of the other cases are virtually the same.

Statements	Reasons
(a) Let $\triangle ABC$ be given. Choose D on $\overleftrightarrow{BC}$ such that $B * C * D$.	(a) --?--
(b) --?--	(b) Proposition 16
(c) $m\angle ACD + m\angle ACB > m\angle ABC + m\angle ACB$	(c) --?--
(d) --?--	(d) Proposition 13
(e) $\pi > m\angle ABC + m\angle ACB$	(e) --?--

Task 21: Is Proposition 17 true in elliptic space?

Task 22: Here is Euclid's Proposition 18. [Note: the word *subtends* is synonymous to the phrase *is opposite to*.]

> Proposition 18: In any triangle, the greater side subtends the greater angle.

Here is Euclid's proof in modern lingo.

98

Statements	Reasons
(a) Let $\triangle ABC$ be given, and suppose $AC > AB$. We must show that $m\angle ABC > m\angle BCA$. Choose D such that $A*D*C$ and $AD = AB$.	(a) --?--
(b) --?--	(b) Definition of exterior angle
(c) $m\angle ADB > m\angle DCB$	(c) --?--
(d) --?--	(d) Proposition 5
(e) $m\angle ABD > m\angle DCB = m\angle BCA$	(e) --?--
(f) --?--	(f) Crossbar Continuity Axiom
(g) $m\angle ABC > m\angle ABD$	(g) --?--
(h) --?--	(h) Transitive Property of Analysis

Task 23: Here is Euclid's Proposition 19.

> Proposition 19: In any triangle, the greater angle subtends the greater side.

And here's Euclid's proof.

Statements	Reasons
(a) Let $\triangle ABC$ be a triangle such that $m\angle B > m\angle C$. We must show that $AC > AB$. For suppose to the contrary, that $AC \leq AB$.	(a) --?--
(b) --?--	(b) Case 1
(c) $m\angle B = m\angle C$	(c) --?--
(d) --?--	(d) RAA Conclusion
(e) Suppose $AC < AB$	(e) --?--
(f) --?--	(f) Proposition 18
(g) Steps a and f are a contradiction	(g) --?--

Task 24: Here is Euclid's Proposition 20, which is the famous Triangle Inequality.

Proposition 20: In any triangle, two sides taken together in any manner are greater than the remaining one.

This proposition is an important one, for it gets to the heart of what we mean when we say the undefined term *line*. Indeed, it is often interpreted as "the closest distance between two points is along a straight line," and the ancient Epicureans believed that this fact is so obvious that it requires no proof. They argued that even an ass knows the Triangle Inequality; for if you put an ass in one corner of a triangular room, and hay in another corner, the ass does not first go to the corner without the hay, and then make its way to the hay. It goes directly to the hay. Nevertheless, Euclid provided a proof.

(a)   Did the Epicureans have a point? Or was Euclid correct in providing a proof? Justify your answer.

(b)   Here's Euclid's proof. Fill in the blanks.

Statements	Reasons
(a) Let $\triangle ABC$ be given. We claim that $AC + AB > BC$. For let D be a point such that $B * A * D$ and $DA = AC$	(a) --?--
(b) --?--	(b) Proposition 5
(c) $m\angle BCD > m\angle ACD$	(c) --?--
(d) --?--	(d) Substitution Property
(e) $DB > BC$	(e) --?--
(f) --?--	(f) Definition of segment measure
(g) $DA + AB > BC$	(g) --?--
(h) --?--	(h) Substitution Prop., steps a and g

(c)   If you create a universe, you can, of course, make up any distance formula you want. (In fact, you need not make up a distance formula at all.) However, the distance formulas that are most beloved by mathematicians are called *metrics*, because universes with metrics (called *metric spaces*) are so much fun. Given any three points,

Universe 8: Complete Neutral Geometry

A, B, and C, a distance formula d is called a *metric* if it has the following four properties:

(i) $d(A, A) = 0$

(ii) $d(A, B) = d(B, A)$

(iii) $d(A, B) \geq 0$

(iv) $d(A, B) \leq d(A, C) + d(C, B)$ (The Triangle Inequality)

Questions: Do *we* live in a metric space? Is Poincaré Disk Space a metric space?

Task 25:   Here's a little Triangle Inequality word problem, for all you Special Relativity fans out there. A thinks that B is going at 88.535% of the speed of light. B thinks that C is going at 66.404% of the speed of light. So C must think that A is going between --?--% of the speed of light and --?--% of the speed of light. Hint: Think of the lovely little triangle that A, B, and C must make on the Poincaré Disk.

Task 26: Here is a modern rendition of Euclid's Proposition 21.

Proposition 21: Let triangle $\triangle ABC$ be given. Let D be a point inside $\triangle ABC$. Then:

(1) $AB + AC > BD + CD$.

(2) $m\angle BDC > m\angle BAC$.

(a)   The proposition talks about point D being inside $\triangle ABC$. Define *inside*. I'll start you off: "A point D is *inside* a triangle $\triangle ABC$ if $\overrightarrow{BA} * \overrightarrow{BD} * \overrightarrow{BC}$, ...

(b)   Here's Euclid's proof! Fill in the blanks.

101

Statements	Reasons
(a) $\overrightarrow{BA} * \overrightarrow{BD} * \overrightarrow{BC}$	(a) --?--
(b) --?--	(b) Crossbar Continuity Axiom
(c) $AB + AE > BE$	(c) --?--
(d) --?--	(d) Addition Property of Analysis
(e) $AE + EC = AC$	(e) --?--
(f) --?--	(f) Substitution Property
(g) $EC + ED > CD$	(g) --?--
(h) --?--	(h) Addition Property of Analysis
(i) $BD + ED = BE$	(i) --?--
(j) --?--	(j) Substitution Property
(k) $AB + AC > CD + BD$, which proves part (i)	(k) --?--
(l) --?--	(l) Proposition 16
(m) $m\angle CEB > m\angle BAC$	(m) --?--
(n) --?--	(n) Transitive Property

Task 27: Let three lines segments, $\overline{AB}$, $\overline{CD}$, and $\overline{EF}$ be given, such that:

$$AB + CD > EF,$$

$$CD + EF > AB, \text{ and}$$

$$AB + EF > CD.$$

Using only a compass and straightedge, construct a triangle of lengths AB, CD, and EF. For a "hint," see Figure 8-4.

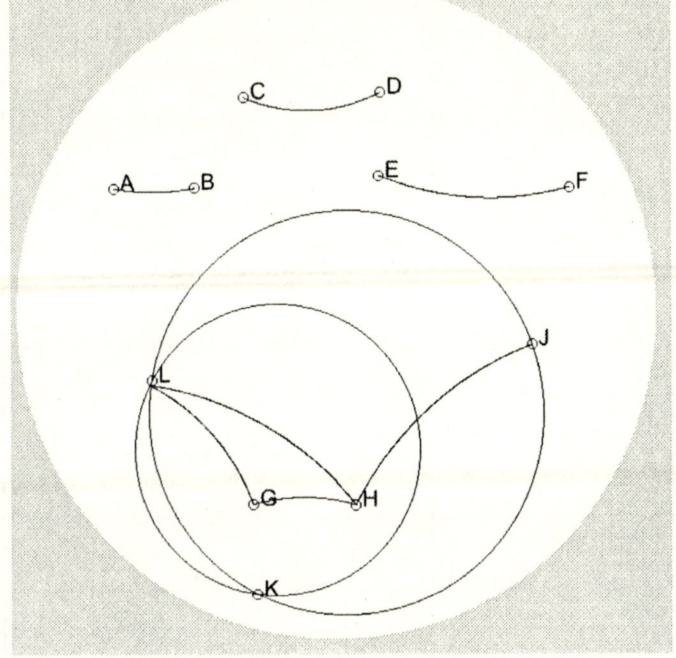

Task 28: Here is a modern rendition of Euclid's Proposition 22.

Proposition 22:   Let three lines segments, $\overline{AB}$, $\overline{CD}$, and $\overline{EF}$ be given, such that:

$$AB + CD > EF,$$

$$CD + EF > AB, \text{ and}$$

$$AB + EF > CD.$$

Then there exists a triangle with sides of lengths AB, CD, and EF respectively.

(a)   How does this proposition compare with Proposition 20?

(b)   Here's Euclid's proof. Fill in the blanks.

Statements	Reasons
(a) WMAWLOGT $AB \leq CD$ and $AB \leq EF$. There exists a circle α with center A and radius CD.	(a) --?--
(b) There exists a circle β --?--.	(b) Set Theory (or Euclid's Postulate 3)
(c) There exists a point G such that $B * A * G$ and $AG = CD$.	(c) --?--
(d) --?--	(d) Definition of circle
(e) $GB = AG + AB$	(e) --?--
(f) --?--	(f) Substitution Property, steps c and e
(g) $CD + AB > EF$	(g) --?--
(h) --?--	(h) Substitution Property
(i) G is outside of circle β.	(i) --?--
(j) There exists a point H such that --?--.	(j) E-2
(k) $AH = CD$	(k) --?--
(l) --?--	(l) Definition of circle
(m) $AH = AB + BH$	(m) --?--
(n) --?--	(n) Substitution Prop steps m and k
(o) $CD - AB = BH$	(o) --?--
(p) --?--	(p) Given
(q) $EF > CD - AB$	(q) --?--
(r) --?--	(r) Substitution Prop steps q and o
(s) H is inside circle β	(s) --?--
(t) --?--	(t) Circular Continuity Axiom
(u) Since I is on α, $AI = CD$	(u) --?--
(v) --?--	(v) Step b

Task 29: Suppose you are given the angle $\angle DCE$ and the line $\overleftrightarrow{AB}$. With a straightedge and compass, construct an angle $\angle LAB$ such that $\angle LAB \cong \angle DCE$. For a "hint," see Figure 8-5.

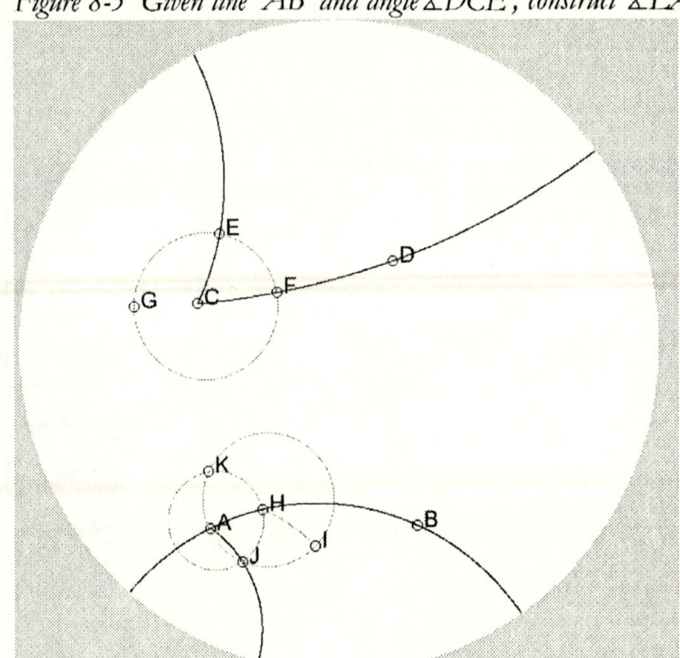

Figure 8-5 *Given line $\overrightarrow{AB}$ and angle $\angle DCE$, construct $\angle LAB \cong \angle DCE$.*

Task 30: Here is a modern rendition of Euclid's Proposition 23.

Proposition 23: Let an angle $\angle DCE$ and a line $\overrightarrow{AB}$ be given.
Then there exists an angle $\angle LAB$ such that $\angle LAB \cong \angle DCE$.

If you're getting a feeling of *déjà vu* as you read this, you're right; we attempted (unsuccessfully) to prove this way back in Universe 3. But this time we have SSS!

(a)   However, before we prove this theorem, we're going to prove a lemma. Up to now, we've been doing pretty much every step in every proof in full detail. This was necessary, because we were laying down the foundations of Neutral Geometry, and we needed to make sure we weren't missing any important axioms. But now, we've earned the right to relax a bit, and only focus on the key steps of each proof. In the following proof, I've "zoomed out" a bit. Good luck!

Lemma: Consider line $\overleftrightarrow{AB}$, and consider circle α with center A and radius CD. Then $\overleftrightarrow{AB}$ intersects α at exactly two points, and these points are on opposite sides of A.

Statements	Reasons
(a) There exists a point E such that $B*A*E$ and $AE = CD$.	(a) --?--
(b) --?--	(b) Definition of circle
(c) There exists a point F such that $E*A*F$ and $AF = CD$.	(c) --?--
(d) --?--	(d) Definition of circle
(e) E and F are the only two points where $\overleftrightarrow{AB}$ intersects α.	(e) --?--

(b)   Use the lemma in the following proof. This proof is not due to Euclid; Euclid's "proof" in *The Elements* has a flaw. Again, I "zoomed out" a bit on this proof, and left out or consolidated some obvious steps. You should be able to handle it.

Statements	Reasons
(a) There exists a circle α with center A and radius CD.	(a) --?--
(b) Let F be the point where --?-- .	(b) Lemma
(c) There exists a circle β with center C and radius CD.	(c) --?--
(d) Let G be the point where --?--.	(d) Lemma
(e) There exists a circle δ with center F and radius GD.	(e) --?--
(f) --?--	(f) Circular Continuity Axiom
(g) $\triangle AFL \cong \triangle CDG$	(g) --?--
(h) --?--	(h) Definition of congruent triangles

Universe 8: Complete Neutral Geometry

Task 31:   You may remember Proposition 24 as one of the two "Hinge Theorems" from high school geometry. On the other hand, maybe you don't. They don't teach geometry like they used to, more's the pity. Why, in *my* day, we had to walk ten miles barefoot through the freezing snow just to prove a lemma... Be that as it may, here is a proof that's *not* due to Euclid. Actually, there's nothing wrong with Euclid's proof, but I like this one better. (For one thing, it's shorter.) Fill in the blanks.

Proposition 24: Let $\triangle ABC$ and $\triangle DEF$ be given, such that $AB = DE$ and $AC = DF$. Suppose $m\angle A > m\angle D$. Then $BC > EF$.

Statements	Reasons
(a) WMAWLOGT $DE \leq DF$. There exists angle $\angle EDG$ such that $\angle EDG \cong \angle BAC$.	(a) --?--
(b) We may assume that $DG = $ --?--.	(b) E-2
(c) $\triangle BAC \cong \triangle EDG$	(c) --?--
(d) --?--	(d) Definition of congruent triangles
(e) There exists a ray $\overrightarrow{DH}$ that is the angle bisector of $\angle FDG$.	(e) --?--
(f) We may assume that H --?--	(f) Crossbar Continuity Theorem
(g) $\triangle FDH \cong \triangle GDH$	(g) --?--
(h) --?--	(h) Definition of congruent triangles
(i) $EG = EH + HG$	(i) --?--
(j) --?--	(j) Substitution Property
(k) $EH + HF > EF$	(k) --?--
(l) --?--	(l) Substitution Property
(m) $BC > EF$	(m) --?--

Universe 8: Complete Neutral Geometry

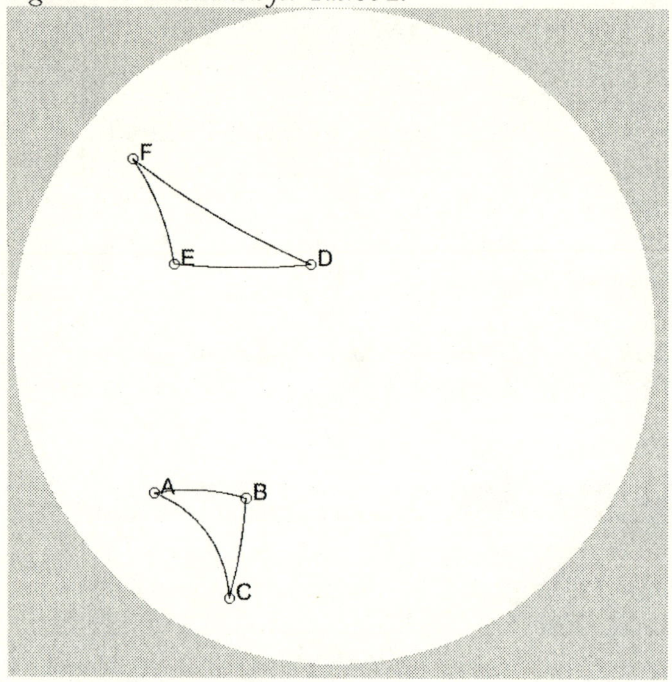

Task 32: See Figure 8-6. A thinks B is going at a velocity of (1, 0°), and that C is going at a velocity of (2, 330°). B thinks that A is going at a velocity of (1, 180°) and C is going at a velocity of (x, 283.2°). Meanwhile, D thinks E is going at a velocity of (1, 180°) and that F is going at a velocity of (2, 151°). E thinks D is going at a velocity of (1, 0°) and that F is going at a velocity of (y, 105.1°). Which is bigger, x or y?

Task 33: Here is the second of the two "Hinge Theorems." This one *is* Euclid's proof. Fill in the blanks.

Proposition 25: Let $\triangle ABC$ and $\triangle DEF$ be given, such that $AB = DE$ and $AC = DF$. Suppose $BC > EF$. Then $m\angle A > m\angle D$.

Statements	Reasons
(a) Suppose, to the contrary, that $m\angle A \leq m\angle D$.	(a) --?--

(b) --?--	(b) Case 1
(c) $\triangle ABC \cong \triangle DEF$	(c) --?--
(d) --?--	(d) Definition of congruent triangles
(e) $BC > EF$	(e) --?--
(f) --?--	(f) RAA Conclusion for Case 1
(g) Now suppose that $m\angle A < m\angle D$.	(g) --?--
(h) --?--	(h) Proposition 24
(i) Steps h and e are a contradiction.	(i) --?--

Task 34: Here is a modern rendition of the Angle-Side-Angle Theorem, Euclid's Proposition 26. You might recall that we already invoked this theorem in Task 11 of Universe 3, in order to prove Proposition 6.

> Proposition 26: (ASA) Let $\triangle ABC$ and $\triangle DEF$ be given, such that $\angle B \cong \angle E$, $\angle C \cong \angle F$, and $BC = EF$. Then $\triangle ABC \cong \triangle DEF$.

Here's Euclid's proof. Enjoy!

Statements	Reasons
(a) We claim that $AB = DE$. For, suppose to the contrary that $AB \neq DE$. WMAWLOGT $AB > DE$.	(a) --?--
(b) --?--	(b) Definition of segment measure
(c) $\triangle GBC \cong \triangle DEF$	(c) --?--
(d) --?--	(d) Definition of congruent triangles
(e) $\overrightarrow{CA} * \overrightarrow{CG} * \overrightarrow{CB}$	(e) --?--
(f) --?--	(f) Definition of angle measure
(g) $m\angle ACB > m\angle DFE$	(g) --?--
(h) --?--	(h) Given

(i) Steps g and h are a contradiction	(i) --?--
(j) --?--	(j) SAS Axiom

Task 35:   "With this proposition begins the second section of the first Book," says Sir Thomas L. Heath, the translator of the Dover Edition of *The Elements*. Euclid's done all of the congruence criteria for triangles, and now his focus will be on parallelism. Of course, this is where the trouble is going to begin again, since *we're* doing Hyperbolic Geometry, and *Euclid* is doing (duh!) Euclidean Geometry. However, so far, it's still good. The next theorem is the famous Alternate Interior Angle Theorem. It states that, given two lines cut by a transversal, if the alternate interior angles are congruent, then the lines are parallel. Here is a modern rendition of Euclid's Proposition 27.

Proposition 27: (Alternate Interior Angle Theorem)

Consider lines $\overleftrightarrow{AB}$ and $\overleftrightarrow{CD}$. Let point E be such that $A * E * B$, and point F be such that $C * F * D$. Assume points A and C are on the same side of $\overleftrightarrow{EF}$.

Suppose $\angle AEF \cong \angle EFD$. Then lines $\overleftrightarrow{AB}$ and $\overleftrightarrow{CD}$ are parallel.

And here's the proof:

Statements	Reasons
(a) Suppose, to the contrary, that $\overleftrightarrow{AB}$ and $\overleftrightarrow{CD}$ are not parallel.	(a) --?--
(b) Then there exists a point G --?--.	(b) Definition of parallel.
(c) $m\angle AEF > m\angle EFG$	(c) --?--
(d) --?--	(d) given
(e) Steps c and d are a contradiction.	(e) --?--

   Universe 8: Complete Neutral Geometry

Task 36: Euclid's Proposition 27 is true in Euclidean universes, hyperbolic universes, and neutral universes, of course. However, it is not true in elliptic universes.

(a) Use the Riemann model of elliptic geometry to give a counterexample to Proposition 27. (If you can find a globe, it will help you to visualize what's going on.)

(b) Where does the proof of Proposition 27 fail for elliptic universes?

Task 37: The next theorem of Euclid's is actually a combination of two famous theorems from high-school geometry. The first states that if corresponding angles are congruent, the given lines are parallel. The second states that if same-side interior angles are supplementary, then the given lines are parallel. Here is a modern rendition of Euclid's Proposition 28.

> Proposition 28: Consider lines $\overleftrightarrow{AB}$ and $\overleftrightarrow{CD}$. Let point G be such that $A * G * B$, and let point H be such that $C * H * D$. Let point E be such that $E * G * H$. Assume A and C are on the same side of $\overleftrightarrow{GH}$.
>
> (a) Suppose $\angle EGB \cong \angle GHD$. Then lines $\overleftrightarrow{AB}$ and $\overleftrightarrow{CD}$ are parallel. (Corresponding Angle Theorem)
>
> (b) Suppose $m\angle BGH + m\angle GHD = \pi$. Then lines $\overleftrightarrow{AB}$ and $\overleftrightarrow{CD}$ are parallel. (Same-Side Interior Angle Theorem)

Here's Euclid's proof. Fill in the blanks.

Statements	Reasons
(a) $\angle EGB \cong \angle AGH$	(a) --?--
(b) --?--	(b) given

Universe 8: Complete Neutral Geometry

(c) $\angle AGH \cong \angle GHD$	(c) --?--
(d) --?--	(d) Proposition 27
(e) $m\angle AGH + m\angle BGH = \pi$	(e) --?--
(f) --?--	(f) given
(g) $m\angle BGH + m\angle GHD = m\angle BGH + m\angle AGH$	(g) --?--
(h) --?--	(h) Addition Property
(i) $\overleftrightarrow{AB} \parallel \overleftrightarrow{CD}$	(i) --?--

Task 38: Is Proposition 28 true in elliptic space?

Universe 9: Hyperbolic Geometry
In Which We Continue Touring Through Euclid's Elements, But With an Attitude That Euclid Would Have Found Bewildering

> *Mathematics is the study of all possible universes.*
> *Physics is merely the study of this universe.*
> --von Neumann

Hyperbolic Parallel Postulate: There exists a line l and a point P not lying on l such that at least two distinct lines parallel to l pass through P.

Hilbert's Axioms of Incidence

IA-1   For every point P and for every point Q not equal to P there exists a unique line l incident with P and Q.

IA-2   For every line l there exists at least two distinct points that are incident with l.

IA-3   There exists three distinct points with the property that no line is incident with all three of them.

Two of Euclid's Postulates

E-2   Given segments $\overline{AB}$ and $\overline{CD}$, there exists a unique point E such that $A * B * E$, and $\overline{CD} \cong \overline{BE}$.

E-4   All right angles are congruent to each other.

Crossbar Continuity Axiom: If $\overrightarrow{AC} * \overrightarrow{AD} * \overrightarrow{AB}$, then ray $\overrightarrow{AD}$ intersects segment $\overline{BC}$. Conversely, if $C * D * B$ and A is not on line $\overleftrightarrow{CB}$, then $\overrightarrow{AC} * \overrightarrow{AD} * \overrightarrow{AB}$.

Circular Continuity Axioms:

(1)   If a circle α with center A has one point inside and one point outside another circle β with center B, then the two circles intersect in exactly two points, and these two points are on opposite sides of $\overline{AB}$.

(2)   If circle α with center A and another circle β with center B intersect at two points, then they intersect at exactly two points, and these two points are on opposite sides of $\overline{AB}$.

Elementary Continuity Axiom:

Suppose a circle with center O intersects a line at point A. If $\overleftrightarrow{OA}$ is perpendicular to the line, then the circle is tangent to the line. Otherwise, the circle intersects the line at exactly two points.

Triangle Congruence Axioms

SAS Axiom: If two sides and the included angle of one triangle are congruent respectively to two sides and the included angle of another triangle, then the two triangles are congruent.

SSS Axiom: If three sides of one triangle are congruent to three sides of another triangle, then the triangles are congruent.

Task 1: Hello! This next stop on our tour of Euclid's *Elements* is a turning point in our geometrical career, because Proposition 29 is the first proposition that's *wrong* in Hyperbolic Space.

> Proposition 29 (Euclidean): Suppose lines $\overleftrightarrow{AB}$ and $\overleftrightarrow{CD}$ are parallel. Let point G be such that $A * G * B$, and let point H be such that $C * H * D$. Let point E be such that $E * G * H$.
>
> Assume A and C are on the same side of $\overleftrightarrow{GH}$. Then the following are all true (in *Euclidean* space, that is):

(a) $\angle AGH \cong \angle GHD$ (converse of the Alternate Interior Angle Theorem)

(b) $\angle EGB \cong \angle GHD$ (converse of the Corresponding Angle Theorem)

(c) $m\angle BGH + m\angle GHD = \pi$ (converse of the Same-Side Interior Angle Theorem)

Proposition 29 is a combination of three theorems from Euclidean Geometry: The Converse of the Corresponding Angle Theorem, The Converse of the Alternating Interior Angle Theorem, and The Converse of the Same-Side Interior Angle Theorem. This proposition states the familiar high-school "facts" that if two parallel lines are cut by a transversal, then any pair of alternate interior angles are congruent, any pair of corresponding angles are congruent, and any pair of same-side interior angles are supplementary. *None* of these "facts" are true in Hyperbolic Space, as Figure 9-1 so eloquently demonstrates. However, there's more to it than that. Each one of these Euclidean theorems is *equivalent* to Euclid's Parallel Postulate. In other words, if any one of them *were* true, Euclid's parallel postulate would have to be true also.

So, we're going to prove that if Proposition 29 *were* true, than Euclid's Parallel Postulate would also be true. Actually, we're only going to prove it for corresponding angles. The proof for alternate interior angles and same-side interior angles is similar, and will be left to the interested reader.

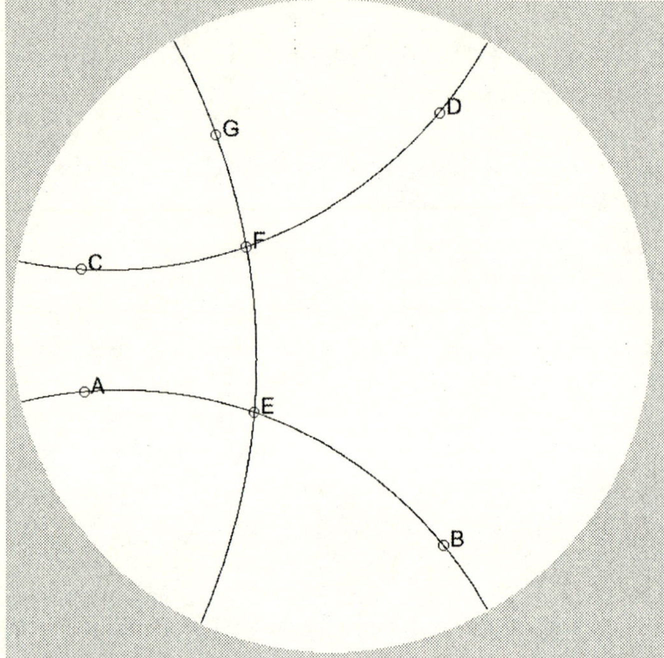

Theorem 9-1: If Proposition 29 is true, then so is Euclid's Parallel Postulate.

Statements	Reasons
(a) Suppose lines $\overleftrightarrow{AB}$ and $\overleftrightarrow{CD}$ are parallel. Let point G be such that $A * G * B$, and let point H be such that $C * H * D$. Let point E be such that $E * G * H$. Assume A and C are on the same side of $\overleftrightarrow{GH}$. Then $\angle EGB \cong \angle GHD$.	(a) --?--

Statements	Reasons
(b) Suppose, to the contrary, that there exists another line $\overleftrightarrow{FG}$ --?--.	(b) RAA Hypothesis
(c) WMAWLOGT F and E are on the same side of $\overleftrightarrow{GH}$, and that $EG = FG$.	(c) --?--
(d) --?--	(d) Proposition 29
(e) $\angle FGB \cong \angle EGB$	(e) --?--
(f) --?--	(f) SAS Axiom
(g) $BE = BF$	(g) --?--
(h) --?--	(h) Proposition 7
(i) Steps g and h are a contradiction	(i) --?--

Task 2: Proposition 30 is also wrong in Hyperbolic Space.

> Proposition 30 (Euclidean): Two lines parallel to a third line are also parallel to each other.

Figure 9-2 shows a counterexample. Proposition 30 is also equivalent to Euclid's Parallel Postulate. In other words, if we wanted to do Euclidean Geometry, we could use Proposition 30 instead of Euclid's Parallel Postulate, and prove all of the same theorems.

> Theorem 9-2: If Proposition 30 is true, then so is the Euclid's Parallel Postulate.

Statements	Reasons
(a) Suppose, to the contrary, that there exists a line l and a point P not on l such that there are at least distinct two lines, m and n, that are parallel to l and pass through P.	(a) --?--
(b) --?--	(b) Proposition 30
(c) Steps a and b are a contradiction.	(c) --?--

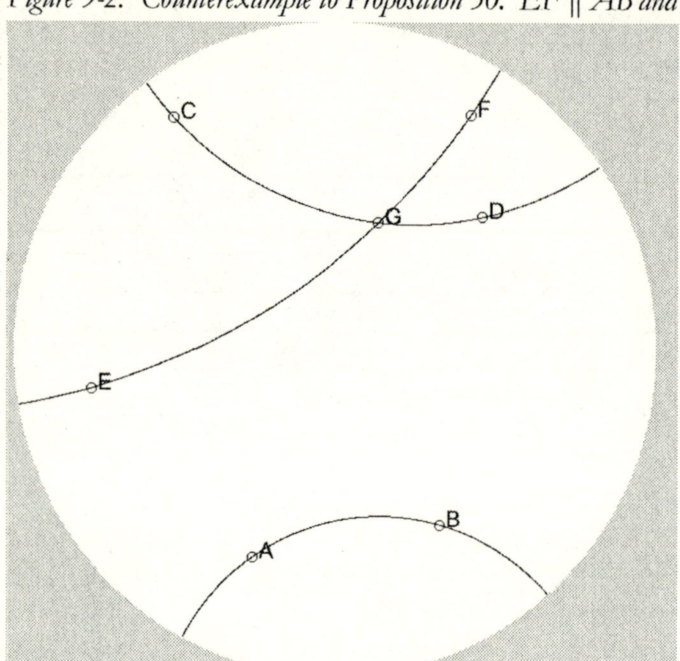

Task 3:   Many scholars have remarked that the next proposition, Proposition 31, is out of order in *The Elements*. It does not depend on any parallel postulate, and thus it is actually a theorem in Neutral Geometry. So it really belongs in Universe 8. Which begs the question: why did Euclid postpone this proof until after Proposition 30? "Presumably," answers Sir Thomas Heath in the Dover Edition, "because he considered it necessary, before giving the construction, to place beyond all doubt that only one such parallel can be drawn." In other words, if he had presented this proposition earlier, it might have begged *another* question: could there by *more* than one line through P parallel to m? Euclid was not ready to go there. In other words, Euclid was afraid of creating non-Euclidean geometry!

> Proposition 31:   Given a line m, and a point P not on m, there exists at least one line through P parallel to m.

Statements	Reasons
(a) There exists two points, B and D, on m.	(a) --?--
(b) --?--	(b) E-2
(c) On $\overleftrightarrow{PD}$, there exists an angle $\angle EPD$ such that $\angle EPD \cong \angle PDC$. WMAWLOGT E and C are on opposite sides of $\overleftrightarrow{PD}$.	(c) --?--
(d) --?--	(d) Proposition 27

Task 4: Proposition 31 is a theorem in Neutral Geometry. Is it a theorem in Elliptic Geometry as well?

Task 5: The next proposition is a combination of two famous theorems from high-school geometry. They are both wrong in Hyperbolic Space. Here is a modern rendition of Euclid's Proposition 32.

Proposition 32 (Euclidean):

(a) The measure of an exterior angle of a triangle is equal to the sum of the measures of the two remote interior angles.

(b) The sum of the measures of the angles of a triangle is equal to the sum of the measures of two right angles.

See Figure 9-3 for a counterexample. As a matter of fact, each part of Proposition 32 is also equivalent to Euclid's parallel postulate. So we now have a total of seven equivalent statements in all, any one of which could serve—along with Neutral Geometry—to create Euclidean Geometry.

This next proof is hard. Indeed, it's the most difficult one we've faced so far. It's both long and subtle. I've shortened it somewhat, and also made it clearer, by skipping obvious steps and by combining algebraic steps; but it's still long. We'll go sweet and slow. At suitable spots, we'll stop for a breather.

Universe 9: Hyperbolic Geometry

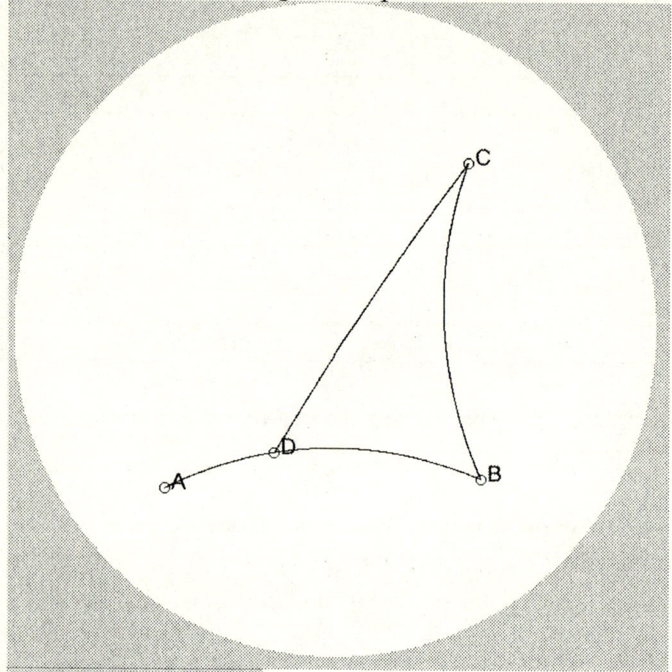

Theorem 9-3: If Proposition 32 is true, then so is Euclid's Parallel Postulate.

Statements	Reasons
(a) Let l be our given line, and P not on l be our given point. There exists point A on l such that $\overleftrightarrow{PA} \perp l$.	(a) --?--
(b) There exists a line $\overleftrightarrow{PB}$ such that --?--.	(b) Proposition 11

	(c) --?--
(c) $\overleftrightarrow{PB} \parallel l$	
(d) Suppose, to the contrary that there is another line $\overleftrightarrow{PC}$ such that --?--. WMAWLOGT $\overrightarrow{PB} * \overrightarrow{PC} * \overrightarrow{PA}$.	(d) RAA Hypothesis
(e) Let E be any point on l that is on the same side of $\overleftrightarrow{PA}$ as B. Then $m\angle AEP + m\angle EPA = 90°$.	(e) --?--
(f) --?--	(f) Step c, and definition of angle measure.
(g) $m\angle BPE = m\angle AEP$	(g) --?--

This is a break in the proof, in order for you to catch your breath, and for you to notice what it is that you just proved. Since E was chosen at random on l, you have just shown that alternate interior angles are always congruent, as long as the transversal goes through P and through a point on l that's on the same side of $\overleftrightarrow{PA}$ as B. Now we'll continue with the proof. The going gets steeper here.

(h) Choose point D_1 on l, on the same side of $\overleftrightarrow{PA}$ as B, such that $AD_1 = PA$. Then $m\angle AD_1P = m\angle D_1PA$. Call this angle measure θ_1.	(h) --?--
(i) $\theta_1 = $ --?--	(i) Proposition 32
(j) $\theta_1 = m\angle BPD_1$	(j) Steps a through g
(k) Choose point D_2 such that $A * D_1 * D_2$ and such that $D_1D_2 = PD_1$. Then $m\angle D_1D_2P = m\angle D_2PD_1 = \theta_2$.	(k) --?--
(l) $\theta_2 = $ --?--	(l) Proposition 32
(m) $\theta_2 = m\angle BPD_2$	(m) steps a through g

Another breather. If you've taken calculus, you probably see where this is heading. We're iterating, and creating an angle θ_n that gets arbitrarily small as n goes to infinity.

    Universe 9: Hyperbolic Geometry

(n) Continuing on in this fashion, $\theta_n = $ --?--	(n) Proposition 32
(o) $\theta_n = m\angle BPD_n$	(o) steps a though g
(p) Since θ_n can be made arbitrarily small as n goes to infinity, choose n such that $\theta_n = m\angle BPD_n < m\angle BPC$. Then $\overrightarrow{PD_n} * \overrightarrow{PC} * \overrightarrow{PA}$.	(p) --?--
(q) --?--	(q) Crossbar Continuity Axiom
(r) Steps q and d are a contradiction.	(r) --?--

Whew! That was quite a proof! But step back, and notice what you did. You proved that for any line l and for any point P not on l, the Euclidean Parallel Postulate holds *if* we assume Postulate 32. Hence, we have the following corollary.

Corollary: There is *no* triangle whose angle sum equals π.

Task 6:   Is Proposition 32 true in Elliptic Space?

Task 7:   So, in Hyperbolic Geometry there is no triangle whose angle sum equals π. Figure 9-3 show a triangle whose angle sum is less than π. But could there also be a triangle whose angle sum is *greater* than π? The answer is no. We know this because of the famous Saccheri-Legendre Theorem.

Saccheri-Legendre Theorem: There is no triangle whose angle sum is greater than π.

Saccheri (1667-1733) and Legendre (1752-1833) both independently discovered this theorem, and they both gave different proofs. (Although Saccheri obviously discovered the theorem first, Legendre was unaware of his work.) However, they both proved it for the same reason. Neither of them was attempting to create non-Euclidean geometry. On the contrary, their goal was to prove

    Universe 9: Hyperbolic Geometry

Euclid's parallel postulate in Neutral Geometry. We now know that this quest was futile; the Poincaré Disk and Klein Disk both follow the postulates of Neutral Geometry, but they're both Hyperbolic, not Euclidean. However, this was way before Poincaré and Klein.

First, they both proved what we proved in Theorem 9-3: that if the angle sum of a triangle is π, then Euclid's Parallel Postulate must be true. So far, so good. Then they thought that if they just proved two more theorems, they would be done. First, they would prove that the angle sum of a triangle is less than or equal to π. Then they would prove that the angle sum of a triangle is greater than or equal to π. Therefore, they reasoned, the angle sum would have to *equal* π, which would prove Euclid's Parallel Postulate.

Their reasoning was impeccable, and they did succeed in proving what we now call the Saccheri-Legendre Theorem: the angle sum of a triangle is less than or equal to π. However, try as they might, they could not prove their second assertion. Many other mathematicians took up the cause, and they all failed miserably. The great Gauss, "the prince of mathematicians," called this assertion "the reef upon which all wrecks occur." He doubted that it could ever be proved, and privately wondered whether Hyperbolic Space was a real possibility. Amazingly, he even wondered whether it were possible that we *live* in a hyperbolic universe (he was ahead of his time!) and apparently did an experiment involving light beams and mirrors on mountaintops—in order to create as big a triangle as feasible—and measured the angles between the light beams to determine whether or not this was so. His experiment was inconclusive, and he recognized that the triangle that he created was probably far too small to get accurate results about the curvature of space. (He was correct about this as well.) Gauss proved many other results in hyperbolic geometry, but he was afraid to publish his work, for fear of ridicule. They were found among his private papers after his death.

So, without further ado, here is a proof of the Saccheri-Legendre Theorem. The proof that we give here is due to Saccheri. It's a long and difficult proof, but after your triumphant experience with the last proof, you should be okay.

Statements	Reasons
(a) Suppose, to the contrary, that there exists a triangle $\triangle ABC$ with angle sum $180° + \varepsilon$, where $\varepsilon > 0$.	(a) --?--
(b) There exists a point D_1 --?--	(b) Proposition 10
(c) There exists a point E_1 such that $AD_1 = D_1E_1$, and $A * D_1 * E_1$.	(c) --?--
(d) --?--	(d) Proposition 15
(e) $\triangle BD_1A \cong \triangle CD_1E_1$	(e) --?--
(f) $\angle CE_1A \cong$ --?--	(f) CPCTC
(g) $\angle BCE_1 \cong \angle CBA$	(g) --?--
(h) $m\angle ACE_1 = $ --?--	(h) Definition of angle measure
(i) $m\angle BAC = m\angle BAE_1 + m\angle E_1AC$	(i) --?--
(j) [angle sum of $\triangle ACE_1$] $= m\angle ACE_1 + m\angle CE_1A + m\angle E_1AC = $ --?--	(j) Substitution Property step h
(k) [angle sum of $\triangle ACE_1$] $= m\angle ACB + m\angle BCE_1 + m\angle E_1AB + m\angle E_1AC$	(k) --?--
(l) [angle sum of $\triangle ACE_1$] $= $ --?--	(l) Substitution Prop, steps i and k
(m) [angle sum of $\triangle ACE_1$] $= m\angle ACB + m\angle CBA + m\angle BAC = $ [angle sum of $\triangle ABC$]	(m) --?--
(n) [angle sum of $\triangle ACE_1$] $= $ --?--	(n) Substitution Property, steps m and a

Time for a breather. At this point, let's be clear about what you *didn't* prove. You did *not* prove that $\triangle ACE_1 \cong \triangle ABC$. And you didn't prove that the corresponding angles are congruent. You simply proved that they have the same angle *sum*. Ready to go on? You are? Good! Let's go!

Statements	Reasons
(o) There exists a point D_2 --?--	(o) Proposition 10
(p) There exists a point E_2 such that $AD_2 = D_2E_2$, and $A * D_2 * E_2$.	(p) --?--

(q) --?--	(q) Proposition 15
(r) $\triangle E_1 D_2 A \cong \triangle C D_2 E_2$	(r) --?--

Ah! We're iterating again. We're doing to $\triangle ACE_1$ exactly what we did to $\triangle ABC$. I won't make you go through all of the substitution steps again. Instead, we'll head straight for the punch line.

(s) [angle sum of $\triangle ACE_2$] = --?--	(s) Same reasoning as in steps f through n.

Of course, we can keep doing this. We can create $\triangle ACE_3$, $\triangle ACE_4$, $\triangle ACE_5$, ... We can create as many triangles as we like, each with the same property.

(t) [angle sum of $\triangle ACE_n$] = --?--	(t) Same reasoning as in steps f through n

So, we've proved that if we're given one triangle with an angle sum bigger than π, then there exists a whole family of triangles with an angle sum bigger than π.

(u) In our original triangle $\triangle ABC$, one of the angles has to be the smallest (or, if the triangle is isosceles, it has to be at least as small as the others). WMAWLOGT the smallest angle is $\angle CAB$. We claim that one of the angles $\angle AE_1 C$ or $\angle CAE_1$ is smaller or equal to half of angle $\angle CAB$. For, suppose to the contrary, that both $m\angle AE_1 C > \frac{1}{2} m\angle CAB$ and $m\angle CAE_1 > \frac{1}{2} m\angle CAB$	(u) --?--
(v) --?--	(v) Addition Property of Analysis
(w) $m\angle BAE_1 + m\angle CAE_1 > m\angle CAB$	(w) --?--
(x) --?--	(x) RAA Conclusion #2

I put "RAA Conclusion #2" in step x, because you may have lost sight of the fact that—starting with step a—we're already in the middle of a proof by contradiction that we haven't concluded yet. So steps u through x were a proof by contradiction within a proof by contradiction... Got that? Now follow your breathing, return to the present moment, and continue with the proof. You're doing great! Just think of how much smarter you're getting, reading and understanding difficult proofs like this one!

(y) WMAWLOGT $\angle CAE_1$ is the smallest of the angles in $\triangle AE_1C$. One of angles $\angle AE_2C$ or $\angle CAE_2$ is smaller than or equal to half of angle --?--	(y) Same reasoning as in steps u through x
(z) So one of the angles in $\triangle AE_1C$ is smaller than or equal to one-fourth of angle $\angle CAB$	(z) --?--
(aa) In fact, in general, one of the angles in $\triangle AE_nC$ is smaller than or equal to --?-- of angle $\angle CAB$.	(aa) Same reasoning as in step z
Basically, we're showing that if we choose a large enough n, we can make an angle as small as we wish.	
(bb) Choose n such that there is an angle in $\triangle AE_nC$ that is smaller than ε. Then the measures of the other two angles in $\triangle AE_nC$ add up to more than 180°.	(bb) --?--
(cc) --?--	(cc) Proposition 17
(dd) Steps bb and cc are a contradiction.	(dd) --?--

Well, that was quite a proof! Congratulations! Go stretch, and perhaps indulge in a refreshing drink.

Now that we've proved our famous theorem, we can combine this with the corollary to Theorem 9-3, to get another corollary.

Corollary: For any triangle, the sum of the angles is less than π.

Task 8: Figure 9-4 shows a counterexample to Proposition 33.

Proposition 33 (Euclidean):   Given quadrilateral $ABCD$. Let $AB = DC$ and let $\overleftrightarrow{AB}$ be parallel to $\overleftrightarrow{DC}$. Then $AD = BC$ and $\overleftrightarrow{AD}$ is parallel to $\overleftrightarrow{BC}$.

Figure 9-4: Counterexample to Proposition 33

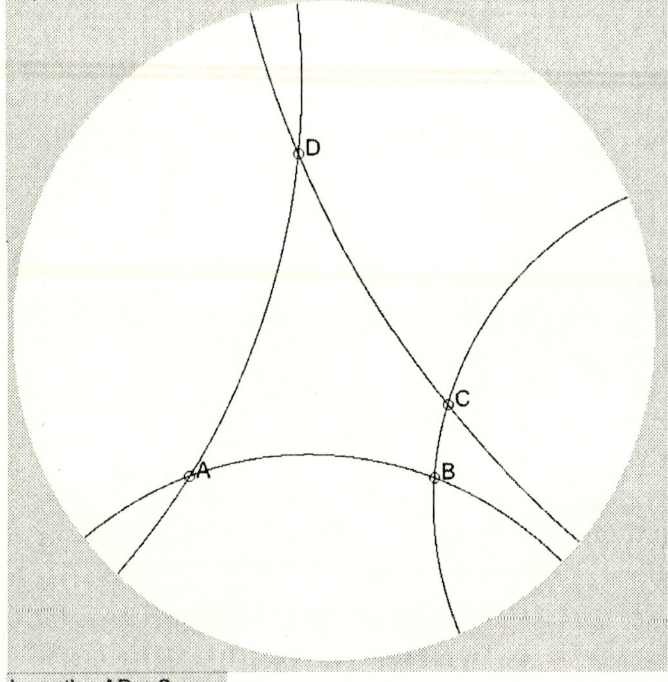

Figure 9-5 shows a counterexample to Proposition 34.

Proposition 34 (Euclidean):   Given parallelogram $ABCD$. Then $AB = DC$, $AD = BC$, $\angle A \cong \angle C$, $\angle B \cong \angle D$, and $\triangle ABD \cong \triangle CBD$.

You probably remember both of these Euclidean theorems from high-school geometry. They're both false in hyperbolic geometry.

   Universe 9: Hyperbolic Geometry

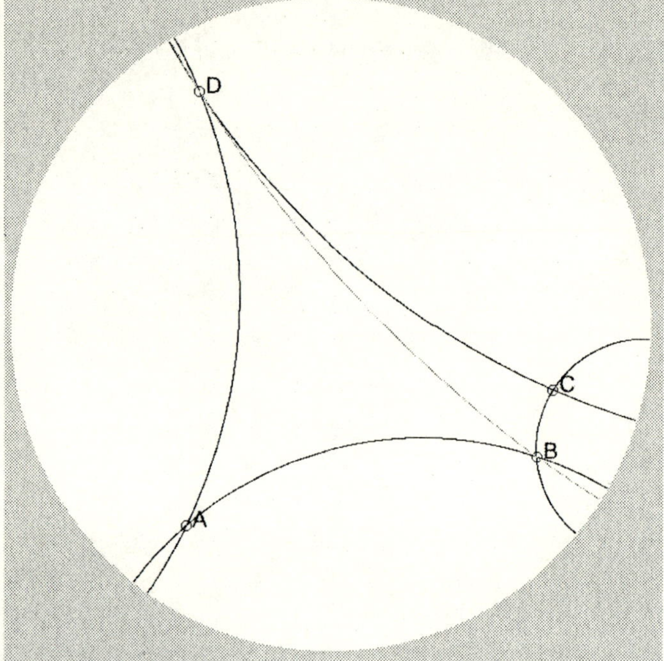

The rest of the propositions, 35 through 48, all deal directly with the concept of *area*, save one. Proposition 46 is the odd man out, so let's take care of this proposition first.

Proposition 46 (Euclidean): Let the segment $\overline{AB}$ be given. Then there exists a square with side $\overline{AB}$.

A *square* is a rectangle with all of the sides congruent. A *rectangle* is a quadrilateral with four right angles. However, rectangles do not exist, so Proposition 46 is not true. Figure 9-6 shows an attempt to construct a rectangle. (In fact, we attempted to construct a square.) Instead, we get a *Saccheri quadrilateral*, which is a quadrilateral whose base angles are right angles and whose base-adjacent sides are congruent. (The base of a Saccheri quadrilateral need not be congruent to the adjacent sides, as it is in Figure 9-6.)

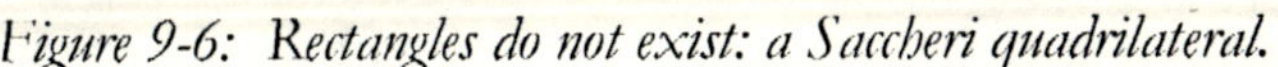

Figure 9-6: Rectangles do not exist: a Saccheri quadrilateral.

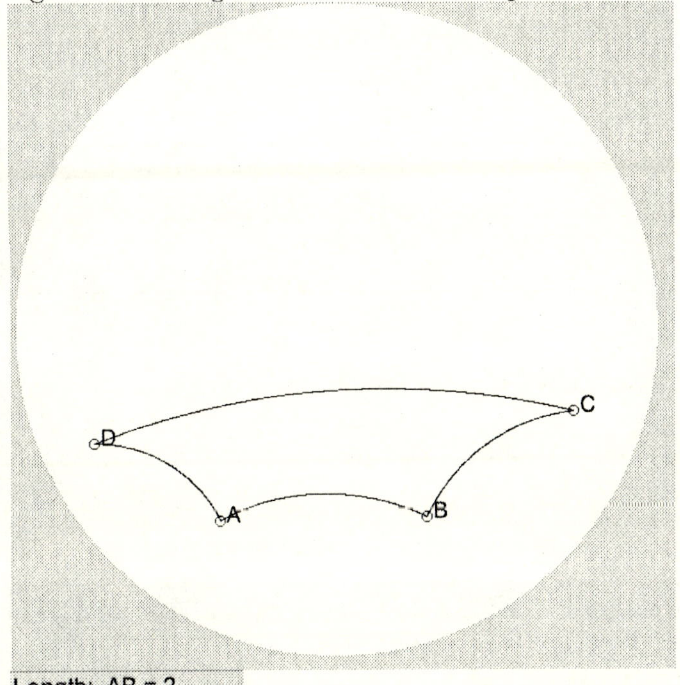

All of this suggests the following theorem:

Theorem 9-4: Rectangles do not exist.

Here is a proof of Theorem 9-4. Fill in the blanks.

Statements	Reasons
(a) Suppose, to the contrary, that there does exist a rectangle $ABCD$.	(a) --?--
(b) --?--	(b) Definition of rectangle
(c) [The angle sum of ΔABC] + [the angle sum of ΔACD] $= 2\pi$	(c) --?--
(d) --?-- $< \pi$	(d) Corollary to the Saccheri-Legendre Theorem
(e) [The angle sum of ΔACD] $< \pi$	(e) --?--
(f) --?--	(f) Steps d and e, and the Addition Property of Analysis
(g) Steps f and c are a contradiction.	(g) --?--

Task 9: In this task, we will prove a famous theorem.

Universal Hyperbolic Theorem: For every line l, and for every point P not on l, there exists infinitely many lines through P parallel to l.

Statements	Reasons
(a) There exists a point Q on l such that $\overleftrightarrow{PQ} \perp l$.	(a) --?--
(b) --?--	(b) Proposition 11
(c) $m \parallel l$	(c) --?--
(d) --?--	(d) IA-2
(e) There exists a line t_1 through R_1 such that $t_1 \perp l$.	(e) --?--
(f) --?--	(f) Proposition 12
(g) $\overleftrightarrow{PS_1} \parallel l$	(g) --?--
(h) --?--	(h) RAA hypothesis

(i) Then quadrilateral PQR_1S_1 is a rectangle.	(i) --?--
(j) --?--	(j) Theorem 9-4
(k) Steps i and j are a contradiction.	(k) --?--

Okay, let's take a breather, and look at what it is that we've done. We have just constructed *two* distinct lines, $\overleftrightarrow{PS_1}$ and m, that go through P and are parallel to l. Already we're not in Euclidean geometry!

(l) --?--	(l) E-2
(m) There exists a line t_2 through R_2 such that $t_2 \perp l$	(m) --?--
(n) --?--	(n) Proposition 12
(o) $\overleftrightarrow{PS_2} \parallel l$	(o) --?--
(p) --?--	(p) RAA hypothesis
(q) Then quadrilateral PQR_2S_2 is a rectangle.	(q) --?--
(r) --?--	(r) Theorem 9-4
(s) Steps q and r are a contradiction	(s) --?--
(t) --?--	(t) RAA hypothesis
(u) Then quadrilateral $S_1R_1R_2S_2$ is a rectangle.	(u) --?--
(v) --?—	(v) Theorem 9-4
(w) Steps u and v are a contradiction	(w) --?--

You probably get the gist. We now have *three* distinct lines, $\overleftrightarrow{PS_1}$, $\overleftrightarrow{PS_2}$, and m, that go through P and are parallel to l. We can now use mathematical induction to complete the proof. We can assume that there exists n distinct lines $\overleftrightarrow{PS_1}$, $\overleftrightarrow{PS_2}$, ..., $\overleftrightarrow{PS_n}$ that go through P and are parallel to l, and then construct yet *another* distinct parallel line, $\overleftrightarrow{PS_{n+1}}$, using the same strategy. Since there are infinitely many points

Universe 9: Hyperbolic Geometry

on line *l* (by E-2), there are infinitely many choices of S_m, and thus infinitely many parallel lines.

Task 10: The rest of Euclid's propositions deal with the concept of *area*.

Euclid never defines what he means by the word *area*, and we're not going to be incredibly hard-nosed about this either. If you ever end up as a math major, you'll take courses in real analysis and measure theory, and discover that the conversation about this topic can become very subtle indeed. For example, consider the function:

$$f(x) = \begin{cases} 1, & \text{if } x \text{ is irrational} \\ 0, & \text{if x is rational} \end{cases}$$

This is certainly a function (it passes the vertical line test!) but what is the "area under the curve" of this function between 0 and 1? Or is this area even defined? Like so many other things in life, the answer depends on whom you ask.

So in this book, we're simply going to concern ourselves with the area of convex polygons. Informally, we can say that the area of a polygon is a positive number associated with that polygon. The scheme for assigning this number need not be unique (for example, we can speak of the area of a polygon as being 4 square feet, 576 square inches, or 34 *glorphs*) but, whatever scheme that we come up with for assigning this number, it must follow the following two rules:

 (a) If two polygons are congruent, then their areas are equal.

 (b) The area of a polygon is the sum of the areas of its non-overlapping parts.

Although we do often measure the area of land in acres, in our everyday world it's more common to measure area in *square* units, such as square feet, square meters, or square miles. And indeed, squares and rectangles are commonly used for *defining* area as well. In calculus, we define the area of a rectangle to be the base times the height, and, using Riemann sums, we define the area of everything else in terms of limits of bunches of rectangles.

However, since squares and rectangles do not even *exist* in hyperbolic space, we've got a problem. Fortunately, there's an easy solution....

Figure 9-7: Area of a triangle: $\partial ABC = 180° - 98.6° = 81.4°$

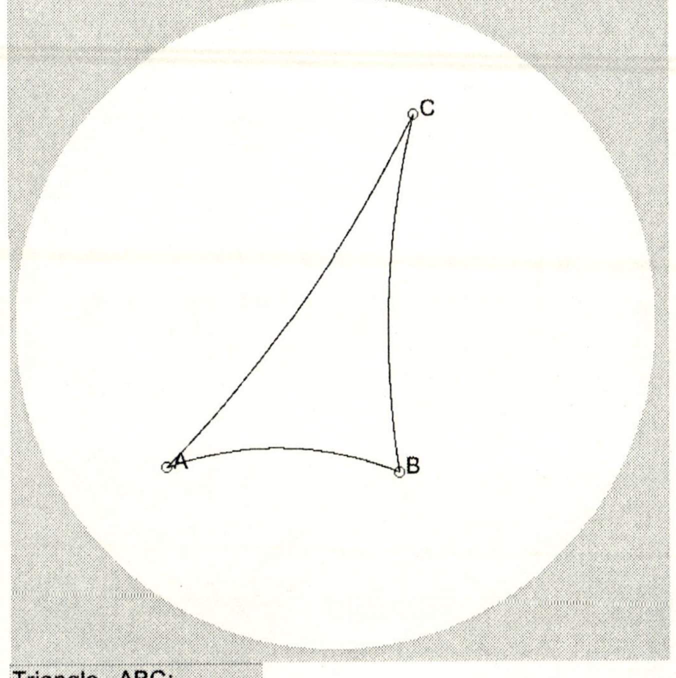

But first, we need to define some terms. Let $\triangle ABC$ be given. We define the *area* of triangle $\triangle ABC$, denoted ∂ABC, to be 180° minus the angle sum. See Figure 9-7 for an illustration. It's easy to check that this scheme for assigning area follows the two rules above. Moreover, since $180° = \pi$, it is feasible in hyperbolic space—amazingly enough—to give the area as a *pure* number, with no units attached. So, for example, the area of the triangle in Figure 9-7 is

133

$407\pi/900$. (Note: in many books on hyperbolic geometry, ∂ABC is called the *defect* of triangle $\triangle ABC$.) Obviously, this scheme for assigning area would never work in Euclidean space, for every triangle would then have zero area.

Figure 9-8: Counterexample to Proposition 35

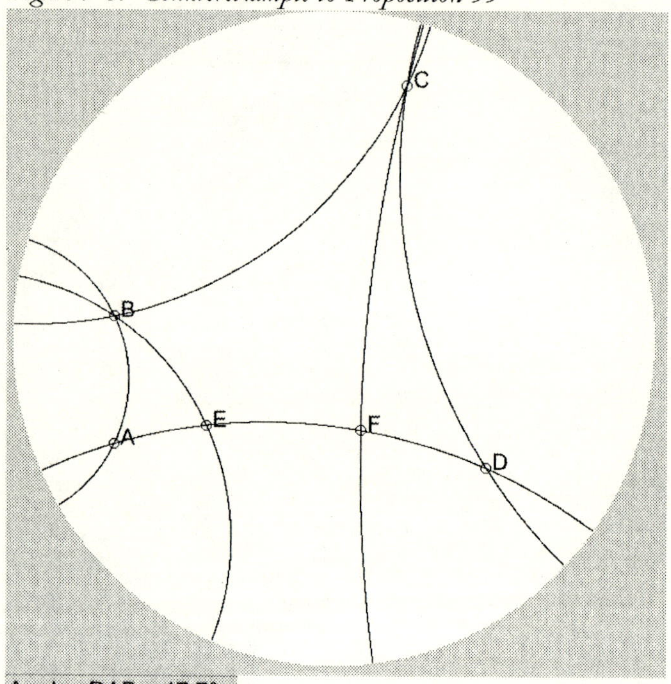

Now we are ready to look at Euclid's next proposition.

> Proposition 35 (Euclidean): Let $ABCD$ and $EBCF$ be parallelograms, such that the four points A, D, E, and F are collinear. (In other words, $\overleftrightarrow{AD} = \overleftrightarrow{EF}$.) Then the two parallelograms have an equal area.

Figure 9-8 shows a counterexample.

(a)   Give the area of a convex quadrilateral in terms of its angle sum.

(b)   Refer to Figure 9-8.   What is the area of parallelogram *ABCD*?

(c)   What is the area of parallelogram *EBCF*?

Figure 9-9: Counterexample to Proposition 36.

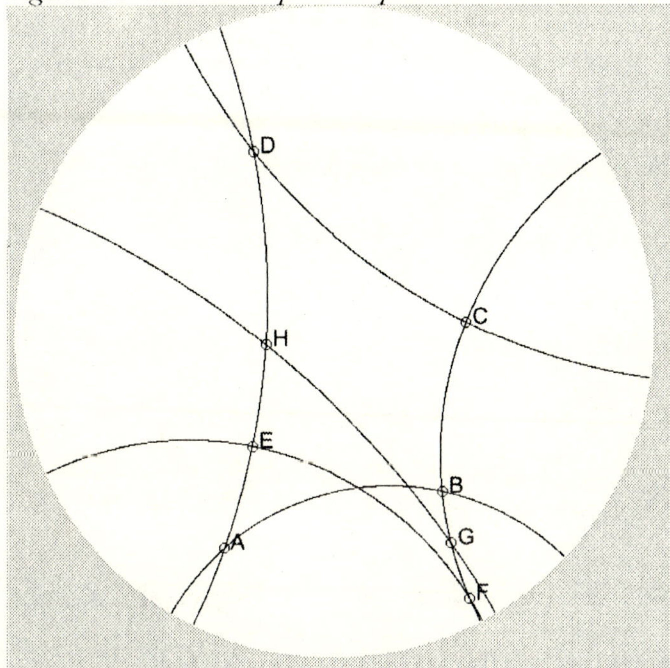

Task 11: Here's Euclid's next postulate.

> Proposition 36 (Euclidean): Let *ABCD* and *EFGH* be parallelograms, such that the four points *A*, *D*, *E*, and *H* are collinear, the four points *B*, *C*, *F*, and *G* are collinear, and *BC* = *FG*. Then the two parallelograms have an equal area.

(a)   Refer to Figure 9-9. What is the area of parallelogram *ABCD*?

(b)   What is the area of parallelogram *EFGH*?

(c)   Suppose somebody gives you a parallelogram. What is its largest possible area?

Task 12:   Here is Euclid's next proposition.

> Proposition 37 (Euclidean): Given triangles $\triangle ABC$ and $\triangle DBC$ such that $\overleftrightarrow{AD}$ is parallel to $\overleftrightarrow{BC}$. Then the two triangles have equal area.

Refer to Figure 9-10 for the counterexample. Find the areas of triangles $\triangle ABC$ and $\triangle DBC$.

Task 13: Here is Euclid's next proposition.

> Proposition 38 (Euclidean): Given triangles $\triangle ABC$ and $\triangle DEF$ such that $\overleftrightarrow{AD}$ is parallel to $\overleftrightarrow{BC}$, the points *B*, *C*, *E*, and *F* are collinear, and *BC* = *EF*. Then the two triangles have equal area.

A counterexample is shown in Figure 9-11. What are the areas of $\triangle ABC$ and $\triangle DEF$?

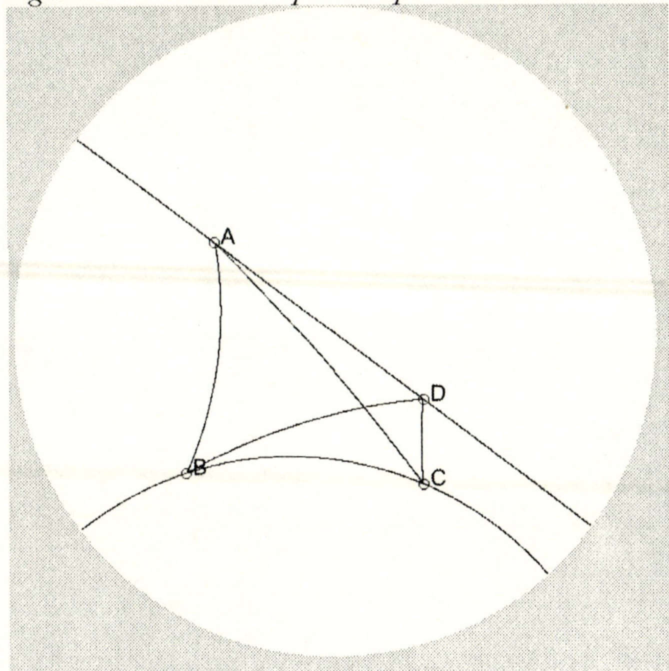

Triangle ABC:
  Length AB = 1.839
  Length AC = 2.206
  Length BC = 1.968
  Angle   A = 35.7°
  Angle   B = 48.2°
  Angle   C = 30.6°
  Angle Sum = 114.5°
Triangle DBC:
  Length DB = 1.863
  Length DC = 0.657
  Length BC = 1.968
  Angle   D = 79.9°
  Angle   B = 11.4°
  Angle   C = 61.9°
  Angle Sum = 153.2°

Figure 9-11: Counterexample to Proposition 38.

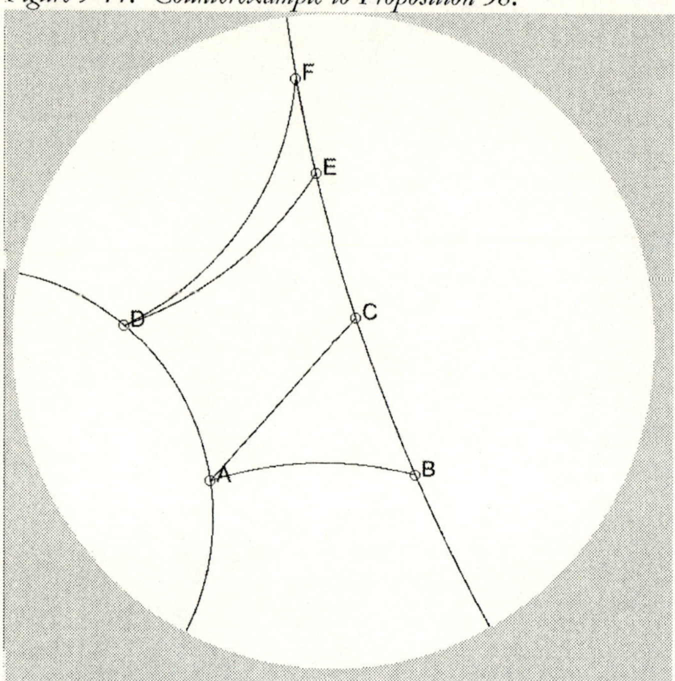

Triangle ABC:
  Length AB = 1.617
  Length AC = 1.515
  Length BC = 1.127
  Angle   A = 30°
  Angle   B = 51.6°
  Angle   C = 61°
  Angle Sum = 142.5°
Triangle DEF:
  Length DE = 1.994
  Length DF = 2.972
  Length EF = 1.127
  Angle   D = 5.9°
  Angle   E = 133.3°
  Angle   F = 15.6°
  Angle Sum = 154.8°

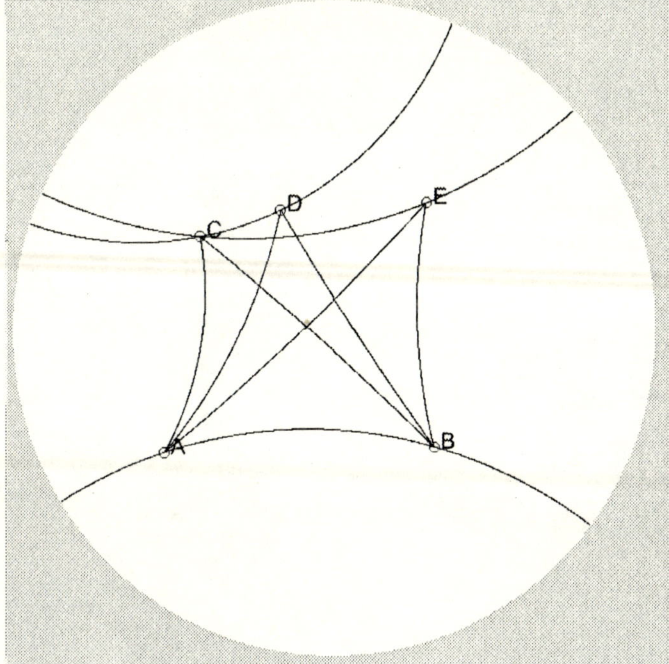

Triangle ABC:
  Length AB = 2.12
  Length AC = 1.792
  Length BC = 2.147
  Angle A = 43.5°
  Angle B = 28.4°
  Angle C = 42°
  Angle Sum = 113.9°
Triangle ABD:
  Length AB = 2.12
  Length AD = 2.01
  Length BD = 1.875
  Angle A = 32.3°
  Angle B = 38°
  Angle D = 43.6°
  Angle Sum = 113.9°
Triangle ABE:
  Length AB = 2.12
  Length AE = 2.582
  Length BE = 1.717
  Angle A = 20.8°
  Angle B = 60.2°
  Angle E = 32.8°
  Angle Sum = 113.9°

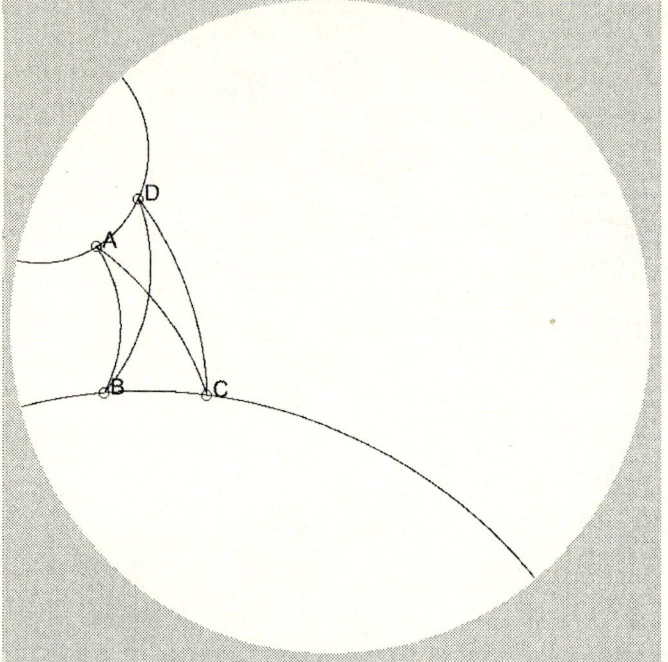

Triangle ABC:
  Length AB = 1.826
  Length AC = 1.766
  Length BC = 1.025
  Angle   A = 21°
  Angle   B = 57°
  Angle   C = 63.3°
  Angle Sum = 141.3°
Triangle DBC:
  Length DB = 2.138
  Length DC = 1.791
  Length BC = 1.025
  Angle   D = 16.7°
  Angle   B = 43.5°
  Angle   C = 81.1°
  Angle Sum = 141.3°

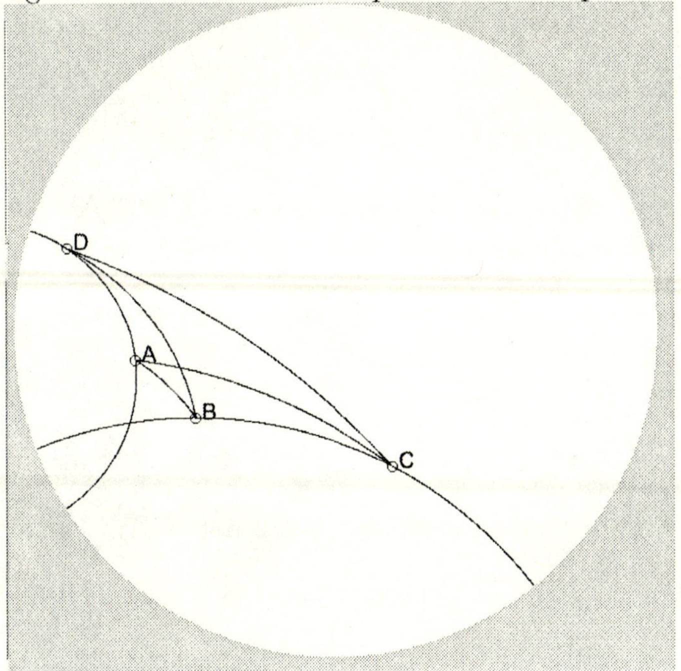

Triangle ABC:
  Length AB = 0.756
  Length AC = 2.101
  Length BC = 1.511
  Angle   A = 25.6°
  Angle   B = 126.3°
  Angle   C = 9.6°
  Angle Sum = 161.4°
Triangle DBC:
  Length DB = 2.462
  Length DC = 3.502
  Length BC = 1.511
  Angle   D = 7.3°
  Angle   B = 101.8°
  Angle   C = 20.1°
  Angle Sum = 129.3°

Task 14:   Euclid's Proposition 39 is a bit vague, and can be interpreted in a couple of different ways. According to Sir Thomas Heath, it reads:

> Equal triangles which are on the same base and on the same side are also in the same parallels.

When Euclid says "equal triangles" here, he means "triangles whose areas are equal." (Sometimes in *The Elements* "equal triangles" means "congruent triangles," so one must always pay attention to the context.) So one way of interpreting Proposition 39 would be as follows.

> Proposition 39 (Euclidean): Given triangles $\triangle CAB$, $\triangle DAB$, and $\triangle EAB$, such that the areas of all three triangles are equal.
>
> Let C, D, and E all be on the same side of $\overleftrightarrow{AB}$. Then points C, D, and E are collinear on a line m, and $\overleftrightarrow{AB}$ is parallel to m.

With this interpretation, Proposition 39 is false in hyperbolic geometry, as Figure 9-12 shows.

And yet, even though C, D, and E are not on "the same parallels" ($\overleftrightarrow{CD}$ and $\overleftrightarrow{CE}$ are different lines in Figure 9-12) nevertheless it *is* true that $\overleftrightarrow{CD}$ and $\overleftrightarrow{CE}$ are both parallel to $\overleftrightarrow{AB}$. So this suggests an alternate version of Proposition 39. See Figure 9-13.

> Alternate Proposition 39: Given triangles $\triangle ABC$ and $\triangle DBC$ such that the areas of both triangles are equal, and such that points A and D are both on the same side of line $\overleftrightarrow{BC}$. Then $\overleftrightarrow{AD}$ is parallel to $\overleftrightarrow{BC}$.

But is Alternate Proposition 39 true in hyperbolic space? For a possible counterexample see Figure 9-14.

Similarly, Euclid's Proposition 40, like Proposition 39, is a bit vague, and admits multiple interpretations. Here's the Sir Thomas Heath translation:

> Equal triangles which are on equal bases and on the same side are also in the same parallels.

So here's one version:

Proposition 40 (Euclidean): Let triangles ΔGAB, ΔHCD, and ΔIEF be given, such that $AB = CD = EF$, such that points A, B, C, D, E, and F are all collinear, and such that points G, H, and I are all on the same side of $\overleftrightarrow{AB}$. Then points G, H, and I are collinear on a line m, and $\overleftrightarrow{AB}$ is parallel to m.

This is again true in Euclidean geometry (Euclid's proof in *The Elements* is fine and dandy) but false in hyperbolic geometry. See Figure 9-15.

Figure 9-15: Counterexample to Proposition 40.

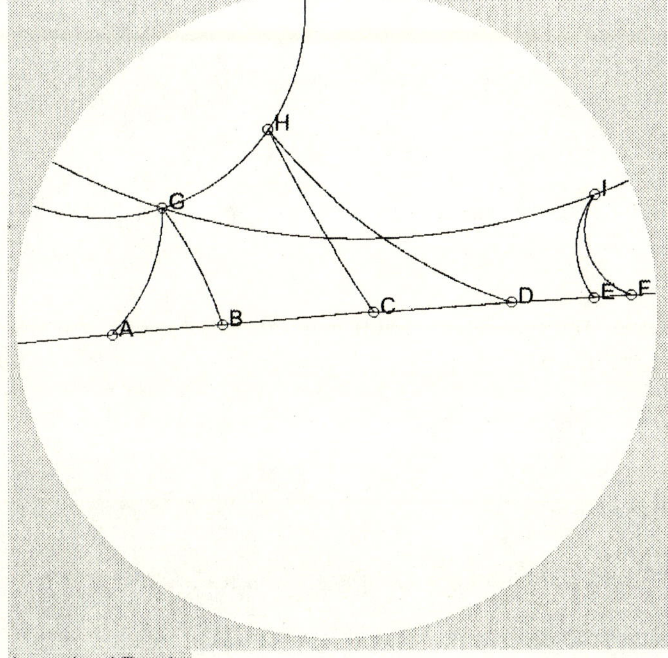

(a)   Explain why Figure 9-14 is not a counterexample for Alternate Proposition 39.

(b)   Since we can't find a counterexample to Alternate Proposition 39, does that mean it's true?

(c)   Give an alternate version of Proposition 40, analogous to our alternate version of Proposition 39. Sketch a picture.

Task 15: Fill in each blank with the word *always*, *sometimes*, or *never*.

(a)   A theorem in Euclidean geometry is —?— true in neutral geometry.

(b)   A theorem in hyperbolic geometry is —?— true in neutral geometry.

(c)   A theorem in neutral geometry is —?— true in hyperbolic geometry.

(d)   A theorem in neutral geometry is —?— true in Euclidean geometry.

(e)   A theorem in Euclidean geometry is —?— true in hyperbolic geometry.

(f)   A statement that's false in neutral geometry is —?— true in Euclidean geometry.

Task 16:   We're going to do Propositions 41 and 43 together as a team, since they are both in the same spirit. Here is Euclid's Proposition 41, in modern lingo:

Proposition 41 (Euclidean): Let parallelogram $ABCD$ and let triangle $\triangle EBC$ be given. Suppose $\overleftrightarrow{AE}$ is parallel to $\overleftrightarrow{BC}$. Then the parallelogram has twice the area of the triangle.

See Figure 9-16 for a counterexample.

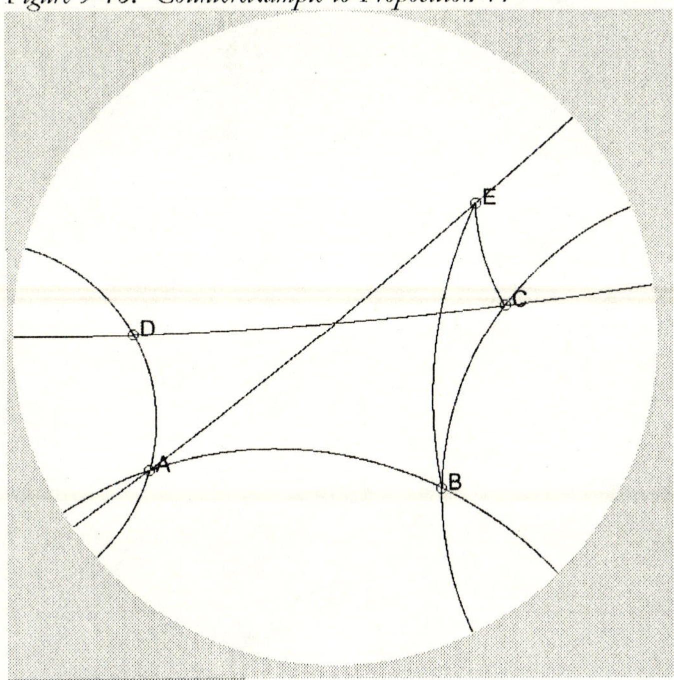

And here is our updated version of Proposition 43:

> Proposition 43 (Euclidean): Let parallelogram $ABCD$ be given. Let E be any given point on the diagonal $\overline{AC}$. Let G be a point on $\overline{AB}$ and F be a point on $\overline{DC}$ such that G, E, and F are collinear, and $\overline{EF}$ is parallel to $\overline{AD}$. Let H be a point on $\overline{AD}$ and I be a point on $\overline{BC}$ such that I, E, and H are collinear, and $\overline{HI}$ is parallel to $\overline{AB}$. Then parallelogram $BGEI$ has the same area as parallelogram $EHDF$.

See Figure 9-17 for a counterexample.

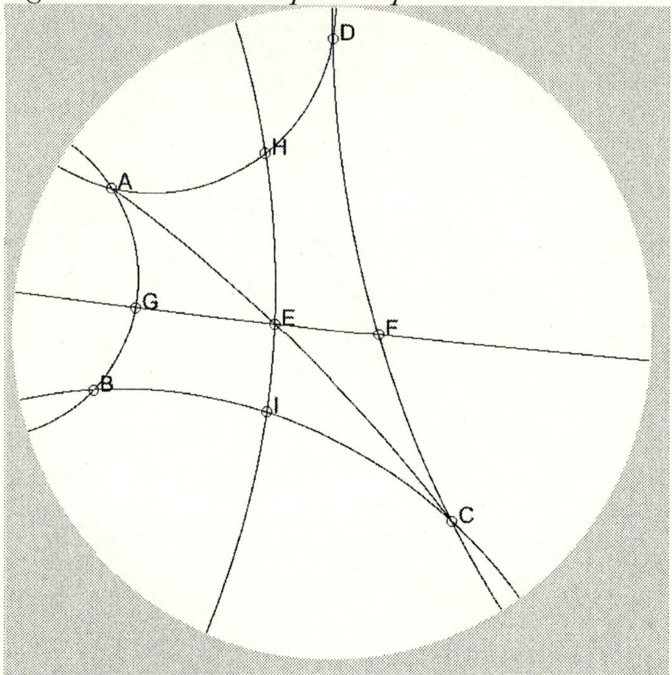

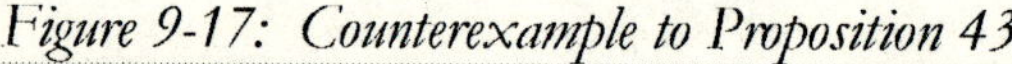

(a)   In Figure 9-16, find the area of parallelogram *ABCD* and the area of ΔEBC.

(b)   In Figure 9-17, find the area of parallelograms *BGEI* and *EHDF*.

Task 17:   Euclid's Propositions 42, 44, and 45 are all recipes for compass and straightedge constructions. None of these constructions work in hyperbolic space; they all implicitly invoke the Euclidean parallel postulate.

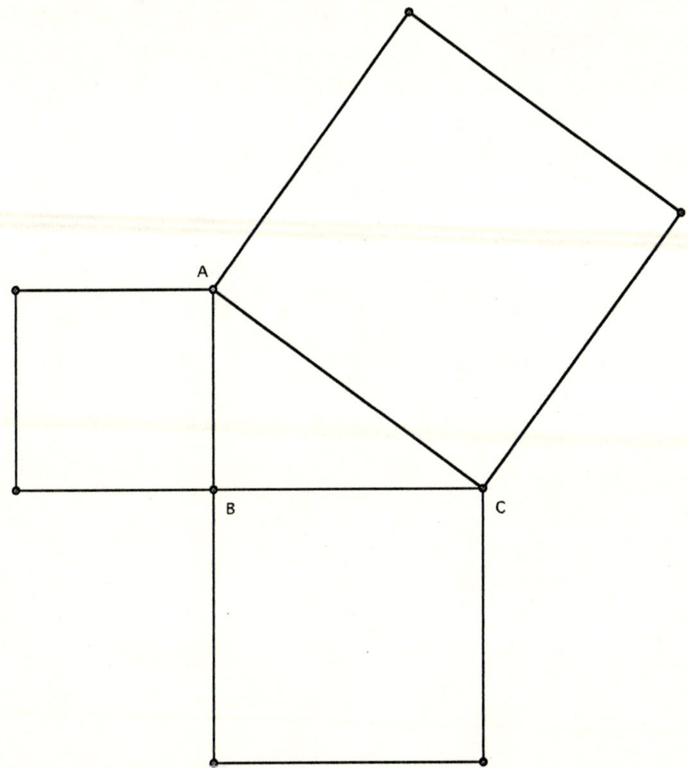

Figure 9-18: In Euclidean space, the area of the square on the hypotenuse is equal to the sum of the areas of the squares on the remaining sides.

Which brings us to the grand finale of Book I of Euclid's *Elements.* Propositions 47 and 48 are among the most famous theorems in all of mathematics.

Proposition 47 (Euclidean): (Theorem of Pythagoras) In a right triangle, the area of the square on the hypotenuse is equal to the sum of the areas of the squares on the other two sides.

Proposition 48 (Euclidean): (Converse of the Theorem of Pythagoras) In a triangle, if the area of the square on one of the sides is equal to the sum of areas of the squares on the other two sides, then the triangle is a right triangle, and the largest square contains the hypotenuse.

This is probably *not* quite the way you remember the Theorem of Pythagoras from high school. Euclid thought in terms of the areas of actual geometrical squares, as in Figure 9-18. This interpretation is certainly nonsensical in hyperbolic space, since squares don't even exist!

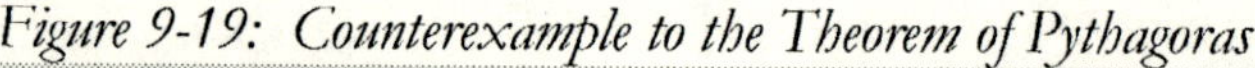

Figure 9-19: Counterexample to the Theorem of Pythagoras

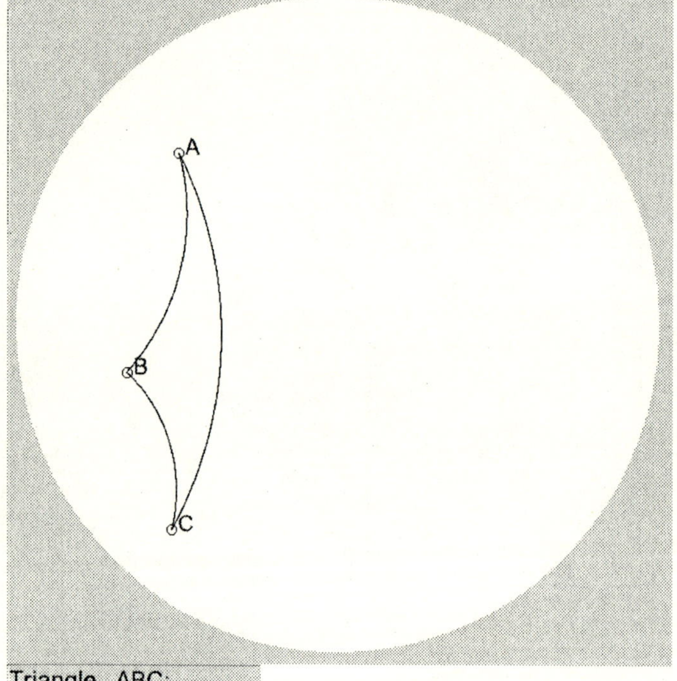

Triangle ABC:
  Length AB = 2.202
  Length AC = 3.539
  Length BC = 2
  Angle   A = 12.2°
  Angle   B = 90°
  Angle   C = 15.1°
  Angle Sum = 117.2°

However, in high school you were probably encouraged to think of this theorem more algebraically than geometrically; "the square of the hypotenuse of a right triangle is equal to the sums of the squares of the remaining sides." But even *this* interpretation does not work in hyperbolic space; see Figures 9-19 and 9-20 for counterexamples.

(a)   Your imaginary friend Harvey says: "In more advanced books in non-Euclidean geometry, it is shown that in hyperbolic space the square of the hypotenuse of a right triangle is always —?— than the sum of the squares of the remaining sides." Unfortunately you missed hearing one of Harvey's words, because there was a clap of thunder. Use Figure 9-19 or 9-20 to deduce the word that you missed.

Figure 9-20: Counterexample to the converse of the Theorem of Pythagoras

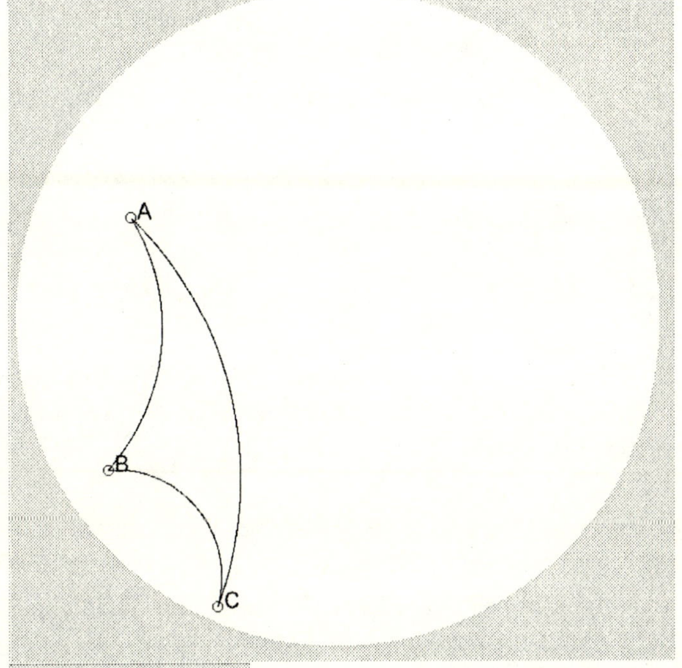

(b)   "In elliptic geometry, the square of the hypotenuse of a right triangle is always —?— than the sum of the squares of the remaining sides." (A globe will help.)

(c)   Just like the elliptic plane can be thought of as the surface of a sphere, the hyperbolic plane can be thought of as the surface of a very strange sphere indeed: a sphere with *imaginary* radius. ("Gee, that's really a big help!" interjects Harvey, sarcastically.) If the sphere happens to have radius i, then a kind of hyperbolic trigonometry can developed. Indeed, given any right triangle $\triangle ABC$, such that $\angle C$ is the right angle, it can be shown that:

$$\sin A = \frac{\sinh a}{\sinh c} \text{ and } \cos A = \frac{\tanh b}{\tanh c}.$$

Using the above formulas, derive the Hyperbolic Theorem of Pythagoras:

$$\cosh c = \left(\cosh a\right)\left(\cosh b\right).$$

[Hint: Recall that $\sin^2 x + \cos^2 x = 1$, and $\cosh^2 x - \sinh^2 x = 1$.]

(d)   (For those who have studied calculus.) Use the formula developed in   part (c), plus the Taylor series for cosh, to show that when $\triangle ABC$ is sufficiently small,

$$c^2 \approx a^2 + b^2.$$

This illustrates that hyperbolic geometry is "locally" Euclidean (just like a sphere is locally planar, or the world is locally flat).

List of Propositions

Here are the Propositions from Book 1 of Euclid's *Elements* in modern lingo.

Propositions 1 – 28 and 31 are all true statements in Neutral Geometry (and thus in both Hyperbolic and Euclidean Geometry). Propositions 4 and 8 we take as axioms; the other ones are theorems.

Propositions 29, 30, and 32 – 48 are false in Hyperbolic Geometry (and thus in Neutral Geometry); however, they are all theorems in Euclidean Geometry.

Proposition 1: Given a segment $\overline{AB}$, there exists a point C such that triangle $\triangle ABC$ is equilateral.

Proposition 2: Let $\overline{BC}$ be a given segment, and let A be a given point not on $\overline{BC}$. Then there exists a point D such that $BC = AD$.

Proposition 3: Given two segments: $\overline{AB}$ and $\overline{CG}$, with $AB > CG$. Then there exists a point E between A and B such that $AE = CG$.

Proposition 4 (SAS Axiom): Given two triangles ABC and DEF. If $AB = DE$ and $AC = DF$, and if $\angle A \cong \angle D$, then the two triangles are congruent.

Proposition 5: (Isosceles Triangle Theorem) In an isosceles triangle, the base angles are congruent.

Proposition 6: (The converse of the Isosceles Triangle Theorem) If two angles in a triangle are congruent, then the opposite sides are congruent (and thus the triangle is isosceles.)

Proposition 7: Let triangle $\triangle ABC$ be given. Then there does not exist a point D (where $D \neq C$) on the same side of $\overline{AB}$ as C such that $AC = AD$ and $BC = BD$.

Proposition 8 (SSS Axiom): Given two triangles $\triangle ABC$ and $\triangle DEF$. If $AB = DE$, $BC = EF$, and $AC = DF$, then $\triangle ABC \cong \triangle DEF$.

Proposition 9: Given an angle $\angle BAC$, there exists a point F such that ray $\overrightarrow{AF}$ is the angle bisector.

Proposition 10: Given segment $\overline{AB}$, there exists a point D between A and B such that $AD = BD$.

Proposition 11: Given a line $\overleftrightarrow{AB}$ and a point C that lies on $\overleftrightarrow{AB}$, there exists a line m through C at right angles to $\overleftrightarrow{AB}$. (We can say "m is *perpendicular* to $\overleftrightarrow{AB}$" or "$m \perp \overleftrightarrow{AB}$".)

Proposition 12: Given a line $\overleftrightarrow{AB}$ and a point C that does not lie on $\overleftrightarrow{AB}$. Then there exists a line through C perpendicular to $\overleftrightarrow{AB}$.

Proposition 13: Given a line $\overleftrightarrow{AB}$ and another line $\overleftrightarrow{CD}$ that intersects $\overleftrightarrow{AB}$ at the point B. We assume that B is between C and D. Consider the angles $\angle ABD$ and $\angle ABC$. Then $m\angle ABD + m\angle ABC = \pi$.

Proposition 14: Let the line $\overleftrightarrow{AB}$ be given. Suppose the points C and D lie on opposite sides of $\overleftrightarrow{AB}$. If the $m\angle ABD + m\angle ABC = \pi$, then the points C, B, and D are collinear.

Proposition 15: (Vertical Angle Theorem) Vertical angles are congruent.

Proposition 16: An exterior angle of a triangle is greater than either of the remote interior angles.

Proposition 17: In any triangle, the sum of the measures of any two angles is less than π.

Proposition 18: In any triangle, the greater side subtends the greater angle.

Proposition 19: In any triangle, the greater angle subtends the greater side.

Proposition 20: (Triangle Inequality) In any triangle, two sides taken together in any manner are greater than the remaining one.

Proposition 21: Let triangle $\triangle ABC$ be given. Let D be a point inside $\triangle ABC$. Then:

$$(1) \quad AB + AC > BD + CD$$
$$(2) \quad m\angle BDC > m\angle BAC.$$

Proposition 22: Let three lines segments, $\overline{AB}, \overline{CD},$ and $\overline{EF}$ be given, such that:

$$AB + CD > EF,$$
$$CD + EF > AB, \text{ and}$$
$$AB + EF > CD.$$

Then there exists a triangle with sides of lengths AB, CD, and EF respectively.

Proposition 23: Let an angle $\angle DCE$ and a line $\overleftrightarrow{AB}$ be given. Then there exists an angle $\angle LAB$ such that $\angle LAB \cong \angle DCE$.

Proposition 24: Let $\triangle ABC$ and $\triangle DEF$ be given, such that $AB = DE$ and $AC = DF$. Suppose $m\angle A > m\angle D$. Then $BC > EF$.

Proposition 25: Let $\triangle ABC$ and $\triangle DEF$ be given, such that $AB = DE$ and $AC = DF$. Suppose $BC > EF$. Then $m\angle A > m\angle D$.

Proposition 26: (ASA Theorem) Let $\triangle ABC$ and $\triangle DEF$ be given, such that $\angle B \cong \angle E$, $\angle C \cong \angle F$, and $BC = EF$. Then $\triangle ABC \cong \triangle DEF$.

Proposition 27: (Alternate Interior Angle Theorem) If two lines are cut by an transversal and a pair of alternate interior angles are congruent, then the lines are parallel.

Proposition 28: Suppose two lines are cut by a transversal.
(a)   If a pair of corresponding angles are congruent, then the lines are parallel. (Corresponding Angle Theorem)
(b)   If a pair of same-side of interior angles are supplementary, then the lines are parallel. (Same-Side Interior Angle Theorem)

Proposition 29 (Euclidean): Suppose a pair of parallel lines are cut by a transversal.
(a)   Then alternate interior angles are congruent. (Converse of the Alternate Interior Angle Theorem)
(b)   Then corresponding angles are congruent (Converse of the Corresponding Angle Theorem)
(c)   Then same-side interior angles are supplementary (Converse of the Same-Side Interior Angle Theorem)

Proposition 30 (Euclidean):   Two lines parallel to a third line are also parallel to each other.

Proposition 31:   Given a line m, and a point P not on m, there exists at least one line through P parallel to m.

Proposition 32 (Euclidean):
(a)   The measure of an exterior angle of a triangle is equal to the sum of the measures of the two remote interior angles.
(b)   The sum of the measures of the angles of a triangle is equal to the sum of the measures of two right angles.

Proposition 33 (Euclidean):   Given quadrilateral $ABCD$. Let $AB = DC$ and let $\overleftrightarrow{AB}$ be parallel to $\overleftrightarrow{DC}$. Then $AD = BC$ and $\overleftrightarrow{AD}$ is parallel to $\overleftrightarrow{BC}$.

Proposition 34 (Euclidean):   Given parallelogram $ABCD$. Then $AB = DC$, $AD = BC$, $\angle A \cong \angle C$, $\angle B \cong \angle D$, and $\triangle ABD \cong \triangle CBD$.

                                          List of Propositions

Proposition 35 (Euclidean): Let $ABCD$ and $EBCF$ be parallelograms, such that the four points A, D, E, and F are collinear. Then the two parallelograms have an equal area.

Proposition 36 (Euclidean): Let $ABCD$ and $EFGH$ be parallelograms, such that the four points A, D, E, and H are collinear, the four points B, C, F, and G are collinear, and $BC = FG$. Then the two parallelograms have an equal area.

Proposition 37 (Euclidean): Given triangles $\triangle ABC$ and $\triangle DBC$ such that $\overleftrightarrow{AD}$ is parallel to $\overleftrightarrow{BC}$. Then the two triangles have equal area.

Proposition 38 (Euclidean): Given triangles $\triangle ABC$ and $\triangle DEF$ such that $\overleftrightarrow{AD}$ is parallel to $\overleftrightarrow{BC}$, the points B, C, E, and F are collinear, and $BC = EF$. Then the two triangles have equal area.

Proposition 39 (Euclidean): Given triangles $\triangle CAB$, $\triangle DAB$, and $\triangle EAB$, such that the areas of all three triangles are equal. Let C, D, and E all be on the same side of $\overleftrightarrow{AB}$. Then points C, D, and E are collinear on a line m, and $\overleftrightarrow{AB}$ is parallel to m.

Proposition 40 (Euclidean): Let triangles $\triangle GAB$, $\triangle HCD$, and $\triangle IEF$ be given, such that $AB = CD = EF$, such that points A, B, C, D, E, and F are all collinear, and such that points G, H, and C are all on the same side of $\overleftrightarrow{AB}$. Then points G, H, and C are collinear on a line m, and $\overleftrightarrow{AB}$ is parallel to m.

Proposition 41 (Euclidean): Let parallelogram $ABCD$ and let triangle $\triangle EBC$ be given. Suppose $\overleftrightarrow{AE}$ is parallel to $\overleftrightarrow{BC}$. Then the parallelogram has twice the area of the triangle.

Proposition 42 (Euclidean): Given a triangle $\triangle ABC$ and given an angle $\angle D$, we can construct with a straightedge and compass a parallelogram that is equal in area to $\triangle ABC$ and has an angle congruent to $\angle D$.

Proposition 43 (Euclidean): Let parallelogram $ABCD$ be given. Let K be any given point on the diagonal $\overline{AC}$. Let E be a point on $\overline{AB}$ and F be a point on $\overline{DC}$ such that E, K, and F are collinear, and $\overline{EF}$ is parallel to $\overline{AD}$. Let H be a point on $\overline{AD}$ and G be a point on $\overline{BC}$ such that G, K, and H are collinear, and $\overline{HG}$ is parallel to $\overline{AB}$. Then parallelogram $BEKG$ has the same area as parallelogram $KHDF$.

Proposition 44 (Euclidean): Let segment $\overline{AB}$, angle $\angle D$, and triangle $\triangle XYZ$ be given. Then we can construct with a straightedge and compass a parallelogram $ABML$ which has an angle congruent to $\angle D$, and an area equal to $\triangle XYZ$.

Proposition 45 (Euclidean): Let quadrilateral $ABCD$ and angle $\angle E$ be given. Then we can construct with a straightedge and compass a parallelogram with an angle congruent to $\angle E$ that has the same area as the quadrilateral $ABCD$.

Proposition 46 (Euclidean): Let the segment $\overline{AB}$ be given. There exists a square with side $\overline{AB}$.

Proposition 47 (Euclidean): (Theorem of Pythagoras) In a right triangle, the area of the square on the hypotenuse is equal to the sum of the areas of the squares on the other two sides.

Proposition 48 (Euclidean): (Converse of the Theorem of Pythagoras) If in a triangle, the area of the square on one of the sides is equal to the sum of areas of the squares on the other two sides, then the triangle is a right triangle, and the largest square contains the hypotenuse.

                                          List of Propositions

Mathematical Symbols Used in This Book

The point A .... A

The line AB .... $\overleftrightarrow{AB}$

The segment AB .... $\overline{AB}$

The measure of segment AB .... AB, or $m\left(\overline{AB}\right)$

The ray AB with vertex A .... $\overrightarrow{AB}$

The angle ABC with vertex B .... $\angle ABC$

The measure of angle ABC .... $m\angle ABC$

The circle α .... α

B is between A and C .... $A * B * C$

Segments AB and CD are perpendicular .... $\overline{AB} \perp \overline{CD}$

Segments AB and CD are congruent .... $\overline{AB} \cong \overline{CD}$

$\overleftrightarrow{AB}$ is parallel to $\overleftrightarrow{CD}$ .... $\overleftrightarrow{AB} \parallel \overleftrightarrow{CD}$

Area of $\triangle ABC$ .... ∂ABC

Printed in the United States
49874LVS00003BA/90